Cultural Commons in the Digital Ecosystem

**Intellectual Technologies Set**

coordinated by
Jean-Max Noyer and Maryse Carmès

Volume 8

# Cultural Commons in the Digital Ecosystem

Maud Pélissier

WILEY

First published 2021 in Great Britain and the United States by ISTE Ltd and John Wiley & Sons, Inc.

ISTE Ltd
27-37 St George's Road
London SW19 4EU
UK

www.iste.co.uk

John Wiley & Sons, Inc.
111 River Street
Hoboken, NJ 07030
USA

www.wiley.com

Library of Congress Control Number: 2021934586

British Library Cataloguing-in-Publication Data
A CIP record for this book is available from the British Library
ISBN 978-1-78630-637-1

# Contents

# Introduction

At the dawn of the 21st century, after having long symbolized the inefficiency of shared property, the commons have reappeared in force as an effective principle of social and cultural struggle against the current dynamics of capitalism. It has imposed itself as the banner of various social movements rising up in the four corners of the world against the extension of private property to all spheres of living together. It is at the heart of the anti-globalization movement, where the activist Naomi Klein, at the Porto Alegre Social Forum in 2001 in a famous text, "Reclaiming the Commons", protested against the privatization of all aspects of life and the transformation of all activity and value into merchandise (natural resources such as education and health). It is regularly found mobilized in environmental movements campaigning for the protection of natural "commons" and the prohibition of access to certain natural resources. The commons are also at the heart of actions to remunicipalize water in certain Italian cities, where the category of common good has been legally recognized alongside that of public property. It thus designates certain categories of resources that must be placed outside the market, managed by a public subject with the participation of citizens. As Pierre Dardot and Christian Laval rightly point out, by uniting disparate forces, the common has gradually become the central category of contemporary anti-capitalism through a resumption of the critique of private property as an absolute condition of social wealth: "What gives meaning to the meeting of these different aspects of the commons in a single designation, is the demand for a new form of 'community' and democratic management of common resources that is more accountable, more sustainable and more just" (Dardot and Laval 2014, p. 97, author's translation).

While the epicenter of the struggle for the commons concerns the protection of common natural resources, it has also spread to the intangibles of the mind. Starting in the 1990s, the strengthening of copyright, the exponential increase in the number of patents filed and the broadening of their scope were perceived as threats to creative dynamics in the artistic, scientific and intellectual fields. Governments and multinational corporations have regularly joined forces to justify the privatization of different forms of knowledge, on the pretext that it was an essential step in the creation of new markets that would bring renewed and sustainable growth.

The emergence and democratization of the Internet as an unprecedented socio-technical ecosystem has opened up a new field of militant struggle for the protection of new forms of commons – sometimes called cultural, information, knowledge or digital – in the face of the threat of disappearance of the creative and sharing dynamics that have been at work since the beginning: "The desire to propose a 'knowledge society' that is shared, equitable and capable of responding to the major crises facing our globalized planet has also mobilized new social actors against the 'knowledge economy', which wants to turn all knowledge into commodities and install financial models in cultural and scientific practices" (Le Crosnier 2015, p. 235, author's translation). Here, the threats of "enclosure" of knowledge come from the desire of certain private companies to extend their field of action or to retain control of their industry. In any case, they are perceived as a potential challenge to the right of universal access to the Internet or the right of access to culture.

Here, as elsewhere, the commons embody a collective will to refound the socio-economic order, based on new regulatory principles and other forms of value and wealth creation. In this respect, these militant movements have made intangible resources, such as free software or the Wikipedia encyclopedia. In different fields, these collaborative communities (Rifkin 2016) are the manifestation of the new conditions of production, circulation and consumption of knowledge made possible by the digital ecosystem. They are new ways of writing, memorizing and reading, which, attached to the process of digitizing the sign, converge towards the promise of seeing the emergence of conditions favorable to collective intelligences (Juanals and Noyer 2010). These commons are also based on a conception of shared ownership and community and decentralized governance, symbolizing a pragmatic utopia embodied by the possibility of seeing a new socio-economic system, a new face of capitalism, unfold in the digital ecosystem.

This book proposes to meet these intellectual movements, most of them militant, which have contributed to the emergence of a real thought on the cultural commons in the digital ecosystem. This path will lead us to identify the different intellectual "places" where this thought has emerged, to identify the socio-economic, technical and political stakes associated with it and finally to highlight the conceptual framework that is proposed for this notion, which is still highly nomadic and polysemic. This exploration is an essential prerequisite to shed light on the foundations of this emerging cultural commons economy.

This thinking of cultural commons in the digital ecosystem is plural. While it is possible to bring about a chronology with pioneering figures, a map with places symbolizing its different manifestations, it is pointless to see it as the expression of one and the same current of thought. This will be the guiding thread of the first part of this book, which will lead us to highlight the importance in the genesis of this thought of the commons of two major intellectual movements. The first movement brings together militant American intellectual figures, most of them university jurists, gathered at the Berkman Center for Internet and Society, and recognized for their battle against the enclosure of the intangible commons of the mind since the end of the 1990s. This movement has had a strong resonance and now has many ramifications, particularly in Europe. The second movement also started in the United States, but in a completely different place: at Indiana University, where the economist Elinor Ostrom, recognized for her research demonstrating the sustainability and effectiveness of land commons, started a research program in the mid-2000s on the question of commons in the field of knowledge. She is credited with helping to open up a research front on the issue of scientific knowledge commons and its articulation with a related but independent movement, open access. Her thinking has also resonated beyond her own Bloomington school, as other scholars in other places have taken up the path of her introductory reflection.

This attempt to reconstruct the various intellectual movements also aims to isolate, in the creative, artistic and scientific fields, the criteria likely to identify resources eligible for the status of cultural commons in the digital ecosystem. Indeed, while this thinking of the commons claims to refound the socio-economic order, it is a question of being able to identify the resources that can claim such a status and then to study the conditions of deployment and survival in complex digital ecosystems.

Thus, in a second part, we will extend this reflection by exploring the conditions of deployment of this cultural commons economy in a specific digital ecosystem, that of the book, relying in each case on the eligibility criteria previously highlighted. The following will be explored in two consecutive chapters, digital library projects as well as common heritage projects, and then self-publishing platforms as illustrations of written commons. At the heart of this empirical analysis is the thorny question of the cohabitation of these cultural commons with the actors of the traditional cultural economy. What guarantees that this emerging cultural commons economy will not be swept away by the force of capitalism's vague "proprietarist" tendencies? Or, conversely, does it not have the intrinsic strength to gradually transform the foundations of cultural capitalism to become the rule rather than the exception? Regardless of the horizon, the cultural commons economy leads to a renewal of the political economy of culture.

# The Intellectual Movement of the Cultural Commons

# Introduction to Part 1

In the cultural field, our investigation into the origins of this notion of the common has led us to identify its birthplace on the other side of the Atlantic. It is indeed there, at the heart of a movement of revolt against certain identified misdeeds of contemporary cultural capitalism, that the notion of the common has been reactivated. This movement was initiated by a group of American jurists gathered at the Berkman Center for Internet and Society (BCIS), founded in 1998 at Harvard University, a unique meeting place for academics and activist experts of the digital world. Among these jurists, all specialists in intellectual property, some, more than others, positioned themselves at the forefront of the scene, such as James Boyle, Yochai Benkler and Lawrence Lessig[1].

The target of their critique was the "proprietarist" evolution of information and cultural markets, symbolizing a drift of the neoliberal economy and a fundamentalist vision of the market. In particular, the evolution of the institutional ecology of cultural markets in the digital ecosystem is, in their view, a major obstacle to free culture, which, after being supplanted throughout the 20th century by a hegemonic commercial popular culture, found a new space for expression. Free creative practices do not fall directly within the scope of copyright, but, for all that, they were quickly condemned by the cultural[2] industries. They symbolize a willingness to share in an ecosystem

---

1 James Boyle, founder of the Center for the Study of the Public Domain, is a professor at Duke Law School in Durham, North Carolina. Yochai Benkler and Lawrence Lessig are professors at Harvard Law School.

2 This chapter is an in-depth version of a French article written in the journal *TIC et Société*, "Communs culturels numériques : origine, fondement et identification", vol. 12, no. 1, 2018.

that facilitates and democratizes popular expression. The notion of the commons was then mobilized by these jurists to account for these transformations. It embodies the possibility of a free cultural economy that is not intended to replace the commercial cultural economy, but rather to find ways of balanced cohabitation.

The notion of the common was not chosen by chance; it has an ancient history. In the economic domain, it has long been disqualified, evoking the subsistence of forms of resource exploitation perceived as an incomprehensible survival of a system deemed inefficient (Guibet Lafaye 2014). It was updated in the 1970s by the economist Elinor Ostrom, who was awarded a Nobel Prize for her work in the field in 2009. On the basis of numerous empirical research works, she has shown that many common natural resources, which were neither managed by the State nor exclusively by the market, constituted an effective yet efficient regulation based on an original model of self-governance and a conception of property in terms of a body of law.

Much later, in the early 2000s, Elinor Ostrom proposed an extension of her approach from land commons to knowledge commons. James Boyle, a jurist with the BCIS, invited her to a conference at Duke University on this topic. On this occasion, in association with librarian Charlotte Hess, Director of the Digital Commons Library at Indiana University[3], she presented a paper entitled "Ideas, Artifacts and Facilities as a Common-Pool Resource" (Hess and Ostrom 2003). They extended these reflections in a book on knowledge commons that paid particular attention to digital archives and libraries as knowledge commons (Hess and Ostrom 2007). This Ostromian intellectual current constitutes the second intellectual locus in which the notion of commons in the digital ecosystem has found a new conceptual life.

Ostrom's intellectual output in this field is not equivalent to what she has produced on the analysis of land commons. However, she has opened up and legitimized the creation of a research program on the theme of knowledge commons in the field of scientific communication, thus offering a theoretical framework for identifying the conditions of their institution and their

---

3 The Digital Library of the Commons is a portal archiving international literature on the commons. All articles are free and open access. This is a collaborative project linked to the *Workshop in Political Theory and Policy Analysis* by Vincent and Elinor Ostrom. See: https://dlc.dlib.indiana.edu/dlc/.

deployment in the digital ecosystem. In France, Ostrom's approach has been extended in a multidisciplinary research program initiated by the economist Benjamin Coriat in 2013. The scope of this program and the interest it shows in the commons in the digital ecosystem, even if it does not focus primarily on the cultural field, deserves to hang around. The rapprochement with the social and solidarity economy also shows their desire to anchor their thinking on communities in a political economy perspective.

# 1

# The Pioneering Approach of Jurists from the Berkman Center for Internet and Society

## 1.1. A critique of the maximalist doctrine of intellectual property

At the end of the 20th century, an ancient debate on the social compromise on which the first copyright legislation was based came back, with force, to the forefront. It originated in the United States. The Berkman Center for Internet and Society (BCIS) was the nerve center of this intellectual battle.

Created in 1998 at Harvard Law School, it brought together lawyers from prestigious universities, specialists in intellectual property and digital law, among whom the most well known are Pamela Samuelson, James Boyle, Julie Cohen, Yochai Benkler and Jonathan Zittrain[1]. Other prominent intellectual figures of the time include constitutional lawyer Lawrence Lessig, libertarian activists such as John Perry Barlow[2], co-founder of the

---

1 Pamela Samuelson is Professor of Law and the Internet at the University of Berkeley. She is recognized for her pioneering work in digital intellectual property law. Julie Cohen, Professor of Law at Georgetown University, specializes in copyright law, Internet regulation and the governance of information and communication networks. She is a member of the Advisory Board of the Electronic Privacy Information Center. James Boyle is Professor of Law and co-founder of the Center for the Study of the Public Domain at Duke University. Jonathan Zittrain is Professor of International and Internet Law at Harvard University. He is a co-founder of the BCIS.

2 John Perry Barlow is also known for his *Declaration of the Independence of Cyberspace*, written at the Davos Forum in 1996 to protest a telecom censorship law signed by Vice President

Electronic Frontier Foundation[3] and entrepreneurs such as Jimmy Wales, founder of Wikipedia. As Anne Bellon points out, within this place "two social universes, law professors and Internet activists, gathered around a critique of the evolution of intellectual property. They contributed to bringing about a counter-discourse that defends the information commons and the value of sharing" (Bellon 2017, p. 166, author's translation).

The exchanges between these two communities, Internet experts and law professors, resulted in alliances around militant actions against the various laws on adapting copyright to the digital age that were passed in the United States in the 1990s. The BCIS, as a venue for the dissemination of theoretical and practical knowledge, is also "a structure of engagement where political discourse mixes with scholarly discourse" (Bellon 2017, p. 181, author's translation). Subsequently, it has become an international model; other similar institutes have been established in different parts of the world. A Global Network of Internet and Society[4] was created in 2012. Today, in France, the CNRS has just formalized the creation of a similar institute: the *Centre Internet et Société* (CIS)[5], considered as a research unit of its own by the *Institut des sciences humaines et sociales*. Mélanie Dulong de Rosnay, director of this center, specializes in issues of digital commons and the link between regulation by law and regulation by technology. She coordinated a research project with one of the members of the BCIS, Juan Carlos de Martin, on the issue of the digital public domain (Dulong de Rosney and de Martin 2012), a theme directly related to the issue of cultural commons, as we will see. Their work is in direct line with the pioneering approach of BCIS jurists.

The intellectual production of these American jurists is consequently on this issue. We will refer here to the writings that we have found most enlightening on the question of the foundations of the political economy of cultural commons. The advantage of their approach is also to translate, on a practical level, into very concrete legal proposals, which have proved to have

Al Gore. He later became an activist in the libertarian cyberculture movement opposed to all forms of state censorship of the Internet.

3 The Electronic Frontier Foundation is an international non-profit organization whose purpose is to promote the protection of freedoms on the Internet. It was founded in 1990.

4 https://networkofcenters.net/.

5 Its objective is to build an interdisciplinary research aiming to shed light on the major technical controversies and the definition of contemporary policies related to digital technology, the Internet and, more broadly, to computer science. See http://cis.cnrs.fr/.

a decisive influence in the evolution of cultural creative practices in the digital ecosystem.

### 1.1.1. *The enclosure of the intangible commons of the mind*

Starting in the 1990s, the cultural and creative industries began to make a significant contribution to economic growth. This has made the US Congress particularly receptive to the arguments put forward by some lobbyists about strengthening intellectual property rights in the face of the rise of an economy where knowledge (in a broad sense) is now seen as the main driving force of long-term economic growth and development[6]. Thus, the prospect of creating new markets justified the ownership of all the intangible forms of knowledge that were at the heart of these new economic valorization processes. According to these jurists, the agreement given to the legalization of the extension of the scope of the patent to the human genome or to the patentability of software were emblematic illustrations of these new commercialization perspectives. For their supporters, the extension of the scope of intellectual property legislation was essential for supporting the new path taken by contemporary capitalism. For their detractors, such as Lawrence Lessig, Yochai Benkler, Pamela Samuelson and James Boyle, these developments in intellectual property legislation in the United States were considered alarming because they called into question the foundations of intellectual property.

These jurists are not opposed to the foundations of the liberal economy and the development of a knowledge-based economy. On the contrary, they are opposed to its excesses and downward spirals that inexorably lead to an "enclosure of the intangible commons of the mind", as James Boyle (2003) put it. Recent developments in intellectual property legislation symbolize what they call a "maximalist teleology", which stems directly from the ideology conveyed by the Washington consensus that markets and exclusive property are the *sine qua non* for economic growth. This growth is based on a rhetoric that equates ownership and economic progress. In this perspective,

6 The OECD defines knowledge-based economies as those that are directly based on the production, distribution and use of knowledge and information. Knowledge-based industries accounted for more than 50% of GDP in the OECD area as a whole at the end of the 1990s, up from 45% in 1985, and are growing faster than GDP in most countries (see data from "Tableau de bord de l'OCDE de la science, de la technologie et de l'industrie 1999 – Mesurer les économies fondées sur le savoir", OECD, Paris).

the granting of exclusive property rights over all forms of culture or knowledge that can be valued in a market is considered the indispensable condition for their efficiency and for economic performance as a whole. However, intellectual property rights are fundamentally different from other property rights.

This approach aims to maintain a non-separation between liberalism and market fanaticism, as they point out, referring to the arguments put forward by the Nobel Prize for Economics winner Joseph Stieglitz. This economist rebels against the dominant idea that there is only one form of capitalism, only one "right" way to manage the knowledge economy. As they were imagined and put in place by the first legislation with Thomas Jefferson, intellectual property rights establish a form of social compromise. In other words, the restriction on the dissemination and use of knowledge (through the establishment of a temporary monopoly situation) is justified as long as it stimulates innovation dynamics and thus promotes growth. This social compromise is a kind of unstable balance that must therefore be preserved. Yet, in the field of knowledge, Stiglitz clearly shows that the beneficial effects of an increasing ownership of knowledge outweigh deleterious effects. They can even lead to a slowing down of innovation[7]. The production of information goods functions on the basis of inputs that are often themselves information. Reinforcing ownership of all these forms of goods therefore also implies reducing access to these inputs and/or increasing their acquisition price, which can have harmful effects on innovation dynamics. This is the observation already made by biologists Michael Heller and Rebecca Eisenberg[8] in the field of biomedical research. Today, this

---

7 Joseph Stiglitz set out this idea in the following article: Grossman, S.J., Stiglitz, J.E. (1980). On the Impossibility of Informationally Efficient Markets. *American Economic Review*, 70, 393–404. He resumed his argument in detail in his book *Making Globalization Work* (2006). Let us quote this eloquent passage: "If patents are made as broad as possible, which is what patent seekers want, there is a real risk of privatizing what is within the public domain, since some (possibly much) of the knowledge covered by the patent is not really 'new'. At least part of what is being patented, and therefore privatized, is knowledge that previously existed – part of common knowledge, or at least of the common knowledge of experts in the area. And yet, once the patent has been granted, the owner can charge others for using that knowledge" (Stiglitz 2006, p. 197).

8 Heller, M.A., Eisenberg, R.S. (1998). Can Patents Deter Innovation? The Anticommons in Biomedical Research. *Science*, May 1. In this article, they show that in a situation where a user needs access to several patented inventions to develop a product, or where intellectual property rights are fragmented among several owners, it can become very costly (in terms of

observation is even shared by economists considered to be orthodox, such as Nobel Prize winner Jean Tirole, who presented an argument similar to Stiglitz's about the ineffectiveness of a measure aimed at lengthening the duration of intellectual property protection on the incentive to create. In particular, in the field of software, he acknowledges that "since the investments have already been made, it is doubtful whether a strengthening of intellectual property has an incentive effect from an economic point of view" (Tirole 2016, p. 567, author's translation).

In the field of culture, the application of this maximalist conception of intellectual property is also detrimental because it leads, in the same way, to a breakup of the original compromise of the copyright law set forth in the American Constitution of 1790, where social progress comes before the protection of the author or inventor, as Lessig repeatedly reminds us. Article I, section 8 of this Constitution affirms, in fact, that "Congress has the power to promote the Progress of Science and useful Arts by securing for limited Times to Authors and Inventors the exclusive Right to their respective Writings and Discoveries" (quoted by Lessig (2004, p. 143)). This social compromise, at the very foundation of the functioning of the markets for intellectual works, is being broken due to a progressive accumulation of legislative measures strengthening the power of rights owners. The particular interest now takes precedence over the general interest. There is thus an inversion in the order of priorities and thus a distancing from the original spirit of the law.

First of all, Lessig blames the exponential increase in the average duration of copyright, which tripled in 30 years at the end of the 20th century. The end of the obligation for all authors to apply for renewal of this right, which was nevertheless written into copyright law from the outset, is also considered to be one of the essential causes of this lengthening: "In 1973, more than 85 percent of copyright owners failed to renew their copyright. That meant that the average term of copyright in 1973 was just 32.2 years. Because of the elimination of the renewal requirement, the average term of copyright is the maximum term" (Lessig 2004, p. 150).

---

transaction costs) for a research team to gather all the necessary authorizations to use the protected resources. Thus, these authors show that in the field of genetics in the United States in particular, this has resulted in an under-use of available resources, hence the idea of the tragedy of the anticommons, with reference to Hardin's paradox, which we will study later.

Following on from this, the scope of copyright has also expanded. Originally covering only certain areas by giving the author the right to "publish" the copyrighted work – a right that was violated if someone decided to republish the work without the author's permission – it gradually extended to derivative rights, giving the author the right to control any "copying" of his or her work. Finally, the limitations restricting procedures have become more flexible and, in particular, the requirement that the work be registered before it can enjoy its protection has been abandoned since the United States ratified the Berne Convention in 1976. Since that date, all intellectual works are *de facto* protected by law.

If all these changes had resulted in socially or environmentally beneficial effects, they could have been justified. But the opposite is happening. In the field of commercial culture, this has led to a strengthening of the already dominant position of the cultural industry majors. Moreover, a large number of these legal developments are the result of intense lobbying by the cultural industries. The example of Disney is often cited as an emblematic illustration of this change. In the 1990s, this company lobbied intensely for an extension of the copyright law by 20 years, because all its favorite characters were going to fall into the public domain at the beginning of the 2000s. This proposal was endorsed in 1998 by the Copyright Term Extension Act (CTEA). It should be recalled that in 1976, when the United States joined the Berne Convention, the copyright term had already been increased to 50 years *post-mortem*.

### 1.1.2. *The threat of disappearance of free culture in cyberspace*

This proprietary spiral does not stop here. The emergence of the Internet has led to a new wave of regulation that is very damaging because it has further restricted what Lessig calls free culture, in other words, all cultural practices that are not regulated by law or authorized by law because they are part of this initial compromise of copyright law. Lessig provides a detailed demonstration of this throughout his work entitled *Free Culture: The Nature and Future of Creativity* (2000). This is a subject he knows well, since he himself defended in the American Court of Justice Eric Eldred, a computer scientist accused of having published works in his digital library, some of which, since the CTEA vote, were again protected by copyright (even though they had previously fallen into the public domain).

The digital ecosystem has radically changed the conditions for the production, distribution and consumption of information and culture. The exchange of digital files, consubstantial to the very existence of this environment, has experienced significant growth with the emergence of peer-to-peer technology, which, it should be remembered, was not originally designed to exchange files protected by copyright, nor to violate the law. Rather than try to take this new situation into account and find a compromise solution, the cultural industries have decided to wage a real war against the promoters and users of this type of platform. Lessig reminds us that the rejection of a new technology by the cultural industries is nothing new, citing the example of the strong reaction of music producers to the appearance of radio cassettes and then CD-ROMs. But this time, he argues that Congress took the copyright owners' side when it should have sought a balance between the interests of each, in accordance with the original spirit of the law.

The Digital Millennium Copyright Act (DMCA), passed in 1998, which legalized the use of technological protection measures (Digital Right Management, DRM) on intangible intellectual works protected by copyright, has also strengthened the control of rights owners over their works: "The DMCA was enacted as a response to copyright owners' first fear about cyberspace. The fear was that copyright control was effectively dead; the response was to find technologies that might compensate" (Lessig 2004, p. 177). However, this solution of legalizing the use of these technological barriers has devastating consequences, because it gives computer code a power to reinforce copyright law far beyond its usual prerogatives.

This DMCA legislation poses a significant risk of reducing the public domain[9] in the digital ecosystem and its contiguous, fair use territories, as Pamela Samuelson points out: "Although not principally aimed at protecting public domain works, the DMCA has significant implications for the digital public domain and for territories contiguous to the public domain" (Samuelson 2003, p. 160). Indeed, since these technical measures constitute private means of regulation because they are controlled by market players (producers and publishers), there is no guarantee that they will scrupulously

---

9 The public domain consists of what is not or no longer protected by intellectual property (copyright or patent law). It includes all information and knowledge that cannot claim protection because it is not considered as works or inventions and all those whose term of protection has expired. The problem is that it has no positive definition from a legal point of view.

comply with the copyright principles protecting the existence of a public domain. Likewise, territories adjacent to the public domain, which are subject to fair use, are also threatened, because technology now allows absolute control of any copy of a protected work in the digital ecosystem. For Lessig, it is the domain of free culture that is under threat, which he prefers to the notion of digital public domain used by Pamela Samuelson. From a legal point of view, this notion was not stabilized and its perimeter was the subject of much discussion. Because of this law (DMCA), "cyberspace", originally a free space protecting anonymity, freedom of expression and individual autonomy, has become a highly regulated space. In its early days, the (computer) code that made "law" on the Internet protected these values. This is no longer the case. The code now has an unprecedented power of regulation, thanks to the evolution of (legal) law, which has allowed it to extend its influence to entire areas of culture that were previously free of control[10].

Understanding the impact of this new legal law on cultural practices is not easy. It requires identifying, upstream, the different uses where copyright is activated, that is, each time a copy of a work is made. There are three possible scenarios:

1) uses not regulated by law because they do not involve copying, such as, for example, the fact of giving or lending a book;

2) uses not regulated by law but which still involve copying. This refers to all uses falling under fair use, or fair dealing, in other words, all uses falling under the exceptions to copyright;

3) uses regulated by law involving copying.

Cases (1) and (2) refer to what Lessig calls free culture and are at the heart of the social compromise of the original copyright law. However, according to him, a large majority of the uses falling under cases (1) and (2) are threatened with illegality in the digital ecosystem because of a "technical anomaly" that implies that all forms of sharing are now based on an act of copying, even in cases that do not fall under copyright law: "Because of this single, arbitrary feature of the design of a digital network, the scope of category 1 changes dramatically" (Lessig 2004, p. 161).

---

10 Lessig develops this argument in his book *Code Is Law and Other Laws of Cyberspace* (Lessig 1999a).

financial
the digital

rst two, as
stakes are
dented ease
ction from
re losing a
is nothing
has resulted
f copyright
violation of
ng networks
the rightful
ing cultural
f music that
l goods are
protected by
lled orphan
the problem

erfeiting, all
therefore to
f sharing to
s how best to
the wrongful

ery enlightening
presentatives of
rulence, such as
nch Minister of
val.... In Cannes
ay is increasing
tion and cultural
thors? A halt to
back to the same
the owner of a
, p. 11, author's

ital book, for example, implies making a copy;
uestion the principle of exhaustion of rights,
ive rights to a copy of a work are extinguished
to possession of it. Lessig gives the example of
digital library. However, he notes that these
ubject to any kind of control by law, are in fact
, by the software used by the publisher, which
bying: "It is code, rather than law that rules"
er words, it is the publishers who now grant
ver thanks to the technique of controlling the
perspective was made possible by the DMCA

native uses considered as unregulated uses.
oming regulated uses, once again, because in
ses imply a copy and therefore a potential
own it. The exception to copyright, fair use,
rmally demand access to the work and its
or's opposition or lack of authorization,
.

now threatened, Lessig takes the example of
are now threatened with illegality in the
rnet, this was, in effect, a totally unregulated
our club onto the Internet, and made it
join, the story would be different. Bots
nd copyright infringement would quickly
81). By detecting all cases of copying of
inevitably facilitates law enforcement and
copy users who have the burden of proving
US law, the boundaries of fair use are not
in France[11]. The limitation to the owner's
e decisions of the courts. The justification
n costs are deemed prohibitive, or social,
is estimated to exceed the owner's loss. In

o-called fair use concerns private copies, parody,
d by the critical, polemical, pedagogical, scientific
ich they are incorporated. For more details, see

the face of such a threat, and in view of the potential for heavy
penalties, creative fair use practices are likely to be curtailed in
ecosystem.

Finally, the uses falling under case (3) are different from the f
they include all uses falling directly under copyright law. The
therefore different. With the emergence of the Internet, the unprece
of sharing cultural works at a lower cost has led to a strong rea
representatives of cultural industries on the pretext that they we
colossal amount of their turnover due to acts of "piracy"[12]. This
new. But, for Lessig, here too, the application of the DMCA law
in a very strong repression that goes far beyond the usual target o
law. However, not all sharing uses on these platforms imply a
the law in the strict sense, that is, the case where users use sharir
as a substitute for purchase, which harms the benefit due to
owner. Indeed, other cases may characterize a practice of shar
content. The user can use the sharing network to get a sample o
he intends to buy later (a rational use considering that cultura
experience goods). He can also use it to have access to content
copyright, but which is no longer sold on the market (so-ca
works). The existence of these different cases inevitably makes
complex, which the DMCA law did not want to see.

With the exception of copying used for commercial count
other forms of sharing are socially beneficial. The challenge is
find a way to remunerate artists and thus allow these forms
survive: "The question we should be asking about file sharing i
preserve its benefits while minimizing (to the extent possible)

---

12 In his book *Du bon usage de la piraterie* (2004), Florent Latrive gives v
illustrations of this battle against "piracy" and the virulent discourse of the r
the cultural industries: "The speeches denouncing copying are now of a rare v
the joint declaration of the Hollywood studios of Jack Valenti and the Fr
Culture of the moment, Jean-Jacques Aillagon, at the 2003 Cannes Film Festi
that day, the aim was to sound the alarm against 'piracy', which every
insidiously on every continent and, by robbing rights owners, threatens crea
diversity.... Who wants the victory of organized crime? The ruin of the a
medical research? No one. The word pirate serves as a scarecrow, it sends
bench of infamy the teenager who downloads a song in MP3 format and
clandestine record duplicating workshop in a Beijing suburb" (Latrive 200
translation).

harm it causes artists" (Lessig 2004, p. 84). However, rather than thinking about a solution of compensation, the DMCA law has allowed for the opposite by giving the copyright owner complete control over all possible uses of his work. The law has sided with the cultural industries and their maximalist view of copyright. Jack Valenti, president of the Motion Picture Association of America (MPAA) since 1966, embodies this new stance of the cultural industries. He has defended before Congress the idea that intellectual property owners should have the same rights and protections as all other property owners in the nation. This is a resurgence of long-standing debates referring to a maximalist vision of literary and artistic property[13].

This maximalist approach to intellectual property reduces free culture to a culture of permission. The existence of a real legal gray area introduced by the DMCA law makes it difficult for everyone to clearly distinguish between what is allowed and what is not, in terms of uses, in the digital ecosystem. In any case, this over-regulation can lead to the curbing of creativity and innovation. This is the main criticism addressed by all these jurists to the current copyright legislation in the digital ecosystem, which they consider inadequate in the face of the technological and anthropological changes underway. Again, they are not opposed to the cultural industries or the copyright legal system. They criticize the over-regulation of recent current laws that only reinforce situations of domination by the majors by providing them with the opportunity to obtain additional income and simultaneously hinder creativity and innovation in the cultural field. Based on this alarming observation, Lessig, like some of his colleagues, called up the notion of the commons to justify the existence of a new threat to culture and knowledge more generally.

## 1.2. The political economy of information commons

The criticism of the maximalist doctrine of intellectual property enunciated by the BCIS jurists as much as their defense of free culture is at the heart of a "political" economy of the information commons. Indeed, they use it as an entry point to justify a renewal of the ethical and anthropological foundations of the information and knowledge economy. While asserting the primacy of individual freedom as a social value, they defend the idea that the privatization and commodification of information, knowledge and culture are

---

13 See Latournerie (2001) for more details.

not the only arrangement defining the horizon of this new economy emerging in the digital ecosystem. The digital ecosystem challenges the usual conditions for the creation, distribution and circulation of information and opens up the possibility of an information economy based on new institutional arrangements known as information commons.

We choose to retain the term "information commons" because it is the term that Benkler and Boyle have used in their respective texts. As we will see, it applies differently to digital resources in the infrastructure, software layer and content areas. In this book, our attention will be focused on information content and, in particular, on resources in the field of works of the mind.

### 1.2.1. *Shared ownership and individual freedom*

It is above all to the jurist Yochai Benkler that we owe the development of the premises of this political economy of the information commons in *The Wealth of Networks*[14]. In this book, he devotes a significant part of his argument to justifying the proximity of his approach to liberal political thought. Let us recall that the latter aims, first and foremost, at preserving and activating individual freedom. However, Benkler emphasizes that precisely the networked information environment increases the capacity and autonomy of individuals by enlarging and diversifying the individual's field of action. It strengthens the possibility for individuals to do things, by and for themselves. This wider range of actions that can be carried out in this environment, alone or together with others, increases the individual space of each person: "The belief that it is possible to make something valuable happen in the world, and the practice of actually acting on that belief, represent a qualitative improvement in the condition of individual freedom" (Benkler 2006, p. 137).

The "user" embodies this new posture in the relationship with the exchange and production of information. He sometimes takes on the role of producer and sometimes that of consumer. In both cases, "they are substantially more engaged participants, both in defining the terms of their productive activity and in defining what they consume and how they consume it" (Benkler 2006, p. 138). This increased space of freedom allows

---

14 This book was published in 2006 and was translated into French in 2009 at the Presses universitaires de Lyon.

each participant to form a more critical judgment and to make his or her own contribution to the evolution of the world. In the cultural space, this translates into the possibility of a new, more transparent and reflective form of popular culture.

This increase in individual freedom made possible by this information commons economy does not imply their adherence to a "rhetoric of the technologically sublime". Technical determinism in the strict sense is based on an erroneous belief. Nevertheless, it is undeniable that digital technologies have an effect on material, social and intellectual living conditions, but this is done differently depending on the institutional ecology in which they are deployed:

> Neither deterministic nor wholly malleable, technology sets some parameters of individual and social action. It can make some actions, relationships, organizations and institutions easier to pursue…. The same technologies of networked computers can be adopted in very different patterns (Benkler 2006, p. 17).

The political economy of the information commons defended in particular by Benkler also introduces new relations between property, market and freedom. Liberal thought considers (exclusive) property as indispensable to the exercise of individual freedom and the market as the institution most likely to bring about an efficient socio-economic order. Benkler questioned these two strong assumptions while maintaining the absolute priority of the principle of individual freedom. Property and the market are not institutional and organizational structures to be considered as natural data. The market must be evaluated in terms of its consequences on the exercise of individual freedom:

> I am offering a liberal political theory, but taking a path that has usually been resisted in that literature – considering economic structure and the limits of the market and its supporting institutions from the perspective of freedom, rather than accepting the market as it is, and defending or criticizing adjustments through the lens of distributive justice (Benkler 2006, p. 16).

Here he joins Lessig's theory, which shows how market regulation can lead to greater control over individual practices and thus over the exercise of freedom. Similarly, property, often presented as a fundamental and natural institution for the efficient functioning of markets, is also revisited. Benkler introduces a different conception of property based on ethics of free sharing whose legitimacy is to preserve and strengthen the exercise of individual freedom. Property and markets should thus be considered as mere domains of human activity with their benefits and limitations: "Their presence enhances freedom along some dimensions, but their institutional requirements can become sources of constraint when they squelch freedom of action in nonmarket contexts" (Benkler 2006, p. 20). Information commons that are based on a conception of shared ownership are thus an essential institutional component of freedom of action in a free society.

Benkler acknowledges that liberal thought has given little importance to culture and has not provided any structured response to the many intellectual criticisms that have arisen throughout the 20th century. In particular, the field of cultural industries has been one of the main axes in the criticism of liberalism as an economic theory, especially in the critical current of the Frankfurt School initiated by Theodor Adorno and Max Horkeimer in the 1920s and which continued throughout the 20th century. The political economy of the information commons provides an answer to these criticisms because it shows how, while remaining within a liberal thought pattern, individuals are able to modify culture by making it more transparent and more inclined to reflection, doubt and questioning. It thus constitutes a response to the latent conflict between the expression of individual freedom and the regulatory framework on which the system of industrial cultural production depends. Everyone comes into the world in a cultural system, embodied in a shared repertoire of traditions and conventions on which social life is based, making the relationship with the other intelligible. For all that, culture is not immutable unconscious data that submits and thus constrains individual freedom. It can become a contested convenience, because it is also the product of a dynamic process of commitment on the part of those who form a culture. In this sense, culture constitutes a framework for negotiation modified by individuals through their communicational relationships. The challenge is to find institutional arrangements that promote the expression of cultural freedom while avoiding an overly hierarchical framework (which controls the interpretation of spaces of meaning) and an overly open framework (which would not allow the recognition of such spaces necessary for mutual intelligibility).

In his book *Remix*, Lessig also stresses the importance of free participatory culture as a means of expressing a fundamental value, freedom of expression: "They reflect upon a capacity for a generation to speak" (Lessig 2008, p. 56). Writing, in the traditional sense of putting words on paper, is the ultimate form of democratic creativity in the sense that everyone has access to the means of writing, which is part of the first fundamental form of learning. The effect on the person producing content can be very positive. Writing on the Web exposes a person to low praise and criticism, as well as the other way around. Writing on this medium means accepting that what you write is subject to debate. Such partisan practices thus encourage an ethics of democracy.

In this perspective, the political economy of the information commons offers a new framework of expression for the deployment of a popular culture that had been "the displacement of folk culture by commercially produced mass popular culture" (Benkler 2006, p. 295) throughout the 20th century. For the critical approach, these new modes of cultural expression stemming from a commercial logic constitute a threat that needs to be fought. For Benkler, property and commons are two institutional and organizational components essential to freedom of action in the cultural field that offer the possibility for everyone to express themselves and create within more complex agencies that are not necessarily reduced to a market logic in the strict sense of the term, but which can also include new forms of social production outside the market: "Their complementary coexistence and relative salience as institutional frameworks for action determine the relative reach of the market and the domain of nonmarket action, both individual or social, in the resources they govern and the activities that depend on access to those resources" (Benkler 2006, p. 24).

The deployment of open cultural practices and information commons does not directly threaten the traditional business economy in the digital ecosystem. What is at stake, however, is the State's ability to make these different arrangements cohabit harmoniously by re-evaluating the institutional rules that condition their operation and their reciprocal relationships. Until now, the State has generally supported, through legislative measures, industrial players in commercial cultural production to the detriment of the creators of the information commons economy. In the current state of affairs, it must be noted that recent legislative measures have led to the reinforcement of existing mono-political situations without any convincing justification in terms of economic efficiency. Moreover, this excess of regulation (through the creation of multiple standards) leads to slowing down creative dynamics and thus innovation processes. The State must change its philosophy of action and become an actor seeking to promote a balance in the deployment of these two cultural spheres. However, this can only be accomplished through minimal regulation ensuring that cultural markets promote the cohabitation of all forms of creativity: "Excessive regulation kills creativity. It stifles innovation. It gives dinosaurs a veto over the future. It squanders the extraordinary potential for democratic creativity offered by digital technology" (Benkler 2006, p. 81). Benkler points out, however, that the State also has a new legitimacy to finance and support social production outside the market, since the repercussions can now be more widely disseminated to increase general welfare.

### 1.2.2. *A new mode of information production*

#### 1.2.2.1. *An information structure conducive to the creation of the commons*

The modes of production and distribution of free cultural resources are deployed within a networked information economy. This third mode of production organization, alongside the market and the company, contributes to the production of information commons in the digital ecosystem.

In the digital ecosystem, the technical conditions for the production and dissemination of information, knowledge and culture have been radically modified. In the 20th century, the cost structure of creating, producing and distributing information resources in markets required significant capital investments that contributed to the development of a highly concentrated economy (which is particularly the case for cultural industries). Now, with digital infrastructure, capital structure for the production and distribution of information, culture and knowledge is decentralized. The physical capital necessary for its overall functioning is now largely held by end-users (with their computers) connected to each other through the Web. This novel decentralized structure thus facilitates the individual or collective production and distribution of these resources, which are no longer reserved for the few owners of capital as in the information and cultural industry of the 20th century. On the other hand, as already mentioned, an important condition for the deployment of such an information economy, and of these new modes of production of commons on the scale of content (cultural, information, knowledge), is the prior existence of an infrastructure (at the physical level) which itself is at least partly common, that is, not entirely privatized.

In this digital environment, the mode of production of the commons is deployed within the framework of a networked information economy that operates on the basis of a new conception of non-exclusive shared ownership. This mode of production corresponds to what Benkler calls a commons-based peer production, the latter being defined as specific institutional forms of structuring rights of access, use and control of resources:

> Commons are specific types of institutional arrangements that govern the use and disposition of resources. Their main characteristic is that no particular person has exclusive control over the use and disposition of a particular resource (Benkler 2003, p. 6).

As he himself mentions, his approach on land commons is close to the original approach of the economist Elinor Ostrom's, who very probably inspired[15] him. However, he rightly emphasizes that the modes of regulation of information commons are not identical to those of land commons. He thus proposes a typology of commons, which, according to him, is based on two distinct criteria: their degree of openness and their mode of regulation. The ocean, the air, the highways, knowledge and culture are open commons: "The most important resource we govern as an open commons, without which humanity could not be conceived, is all of pre-20th century, and much of contemporary science and academic learning" (Benkler 2003, p. 7). Conversely, the land commons described by Ostrom are in limited access to a clearly defined group. The second parameter is the degree of regulation of the commons. Ostrom's land commons are regulated by more or less elaborate rules (some formal, others stemming from social conventions). The information commons deployed in the digital ecosystem vary widely in their degree of regulation. Some are not governed by any rule as resources in the public domain. He calls them open access commons. Others, as we shall see, are based on the existence of rules and social norms of regulation.

---

15 Let us recall that it was in 2003 that Benkler wrote for the first time on the political economy of the information commons, in a special issue of a collection of articles following the Duke colloquium initiated by James Boyle and with Elinor Ostrom as the guest of honor.

## 1.2.2.2. *Economic value and openness of information commons*

Carol Rose (1986), Associate Legal Officer at the BCIS, provided a very enlightening analysis dedicated to the question of the value generated by knowledge and cultural commons. Her argument is to defend the hypothesis that the value of these information resources is correlated with their degree of openness: the more open they are, the more they are used and the more their value increases because they will be a source of positive externalities. Shared resources can create, depending on the context, more wealth and opportunities for society than if they were privately owned. Decreasing control over these resources implies a proportional increase in their social as well as economic value. This principle runs counter to one of the traditional pillars of the knowledge economy, which bases the efficiency of information markets on privatization and maximum control over its resources. In this respect, Carol Rose refers to the work of economist Richard Posner as one of the instigators of such a posture[16].

The decentralized production of the commons contributes to the generation of social value. which can then be reappropriated with a view to economic valorization through innovation processes in the networked information economy. This idea is particularly present in Lessig's work: "Commons also produce something of value. They are resources for decentralized innovation" (Lessig 2001, p. 99). In the field of works of the mind, free cultural practices are new sources of creativity and, as such, they are a potential lever for innovation dynamics that create economic value. In particular, he insists on the fact that actors in the industrial information economy can then exploit this dynamic of social value creation. Moreover, Lessig asserts that his plea for free culture is synonymous with a defense of a free market: "The charge I've been making about the regulation of culture is the same charge free marketers make about regulating markers" (Lessig 2009, p. 76). Over-regulation harms creativity and stifles innovation; "it wastes extraordinary opportunity for a democratic activity that digital technology enables" (Lessig 2004, p. 222). Entrepreneurs who want to innovate in this space cannot do so safely. It is therefore essential to be able to break the chains that hold back these new creative spaces.

---

16 "Exclusive private property is thought to foster the well-being of the community, giving its members a medium in which resources are used, conserved and exchanged to their greatest advantage. There is nothing new about this set of ideas; Richard Posner, a modern-day proponent of neoclassical economics, remarks that the wealth enhancing value of property rights has been well known for several hundred years" (Rose 1986, p. 711).

### 1.2.2.3. *An unprecedented collaborative production mode*

The production modes of information commons do not only follow individual logics, a certain number of which are based on original forms of collaborative organization. Open source software is the archetypal example: "The best-known examples of commons-based peer production are the tens of thousands of successful free software projects that have come to occupy the software development market" (Benkler and Nissenbaum 2006, p. 395). But other information sources are produced according to a similar logic of collaborative organization.

In the digital ecosystem, a growing number of projects to produce information commons is based on collaboration between actors who decide to pool their time, experience and creativity. In the scientific field, Benkler gives the example of two projects: SETI@home, an experimental scientific project using the power of connected computers – 4.5 million users from 226 countries – to catalog astronomical radio signals with a view to identifying extraterrestrial intelligence, and Clickworkers, a project initiated by NASA enlisting tens of thousands of individuals to identify Martian craters. In the field of content creation, the Wikipedia encyclopedia, the Dladshot platform (technological newsletter) or the Gutenberg project (digital library project) are also cited as emblematic illustrations.

All these production modes of information commons are part of what can be called the non-market social economy for several reasons. On the one hand, the producers of these resources are not necessarily professionals and, in any case, they are not financially compensated for their respective contributions. On the other hand, the production of such information commons is not oriented towards the search for profit by the designers of the platforms that host these voluntary contributions. In legal terms, these contributions often fall within the field of associations. The donation is the preferred form of remuneration.

### 1.2.2.4. *An expansion of the mode of production of information commons*

How can such production be the dominant organization of the networked information economy? We can indeed be led to suppose that because of the voluntary nature of these productions, individual contributions are random and are destined to be confined to the margins of information production systems. Benkler devotes a significant part of his writings to justifying the relevance and efficiency of this type of production. The digital ecosystem promotes the activation of social behaviors that, from the peripheries, progressively dominate the networked information economy.

The figure of *homo oeconomicus*, at the heart of the dominant paradigm in economics since Adam Smith, which is that of an individual whose action is entirely oriented towards the search for his individual interest (profit, personal satisfaction, etc.), is called into question by the existence of such social behaviors within the networked information economy. Debates have taken place on the decisive elements of social behavior in the world of free software. For some (Lerner and Tirole 2002), social behaviors can be brought back to the logic of *homo oeconomicus* because even if volunteer developers are motivated by access to a position or social recognition, this indirectly translates the search for an economic gain that can serve as a lever for obtaining future employment. Conversely, others (Lakhani and Wolf 2005) show on an empirical basis that their action is motivated by the search for pleasure through intellectual stimulation, the improvement of their skills or the feeling of being part of a community. Benkler is clearly in favor of these results.

In the field of software, the fact that a growing number of developers are now being paid by their companies to participate in the production of free software does not call into question the foundations of such a collaborative system, which is therefore part of a hybrid economy. On the one hand, the latter have a high degree of autonomy and significant room to maneuver in the management of their activities. On the other hand, this cohabitation remains possible "as long as the principle of reward at the basis of such production, which is not monetary but intangible – the respect and admiration of one's peers, the power to influence the development of a project – remains intact for unpaid developers" (Benkler 2006, pp. 185–186). However, although much importance is attached to this first argument, it alone cannot account for the effectiveness and sustainability of this type of social production in the digital ecosystem. Indeed, this philosophical debate on the decisive elements of social behavior invariably restricts this question to an individual motivation dimension.

There are also other factors that play an equally important role in the growth of these information commons, which are as much a matter of the properties of the digital ecosystem as of the mode of governance of these social productions. Indeed, regardless of the nature and extent of the contributors' social motivations, everyone must work to earn money and provide for their basic needs at the very least. The time that each person devotes to it being thus necessarily reduced in the face of these unavoidable economic constraints, this constitutes an undeniable brake on these dynamics of social production. However, it has been shown that the organization of collaborative production is based on two characteristics that play a decisive role in explaining their growth: modularity and granularity. Indeed, the social production of such information commons takes the form of a sum of independent modules allowing the different contributors to choose autonomously the nature and extent of their participation according to their desires and available leisure time. This mode of social production thus relies on the cooperation of a large number of contributors, each acting according to their possibilities and desires. Moreover, it is based on production modules whose granularity, that is, the individual time and effort that must be invested to produce them, is low. This second organizing property of social production is essential because, as Benkler notes, conversely, "if the finest-grained contributions are relatively large and would require a large investment of time and effort, the universe of potential contributions decreases" (Benkler 2006, p. 101). While these properties of granularity and modularity are common to open-source software, they are also at the heart of how Wikipedia works. Conversely, Benkler cites the failure of the Wikibooks collaborative textbook production project, also designed by Jimmy Wales:

> Very few texts there have reached sufficient maturity to the extent that they could be usable as a partial textbook, and those few that have were largely written by one individual with minor contributions from others... The minimum time requirement required of each contributor is therefore large, and has led many of those who volunteered initially to not complete their contributions (Benkler 2006, p. 101).

The second property that explains the deployment of information commons in the digital ecosystem concerns the mode of governance of these resources. In these modes of social production, collaborative production goes far beyond the simple aggregation of individual contributions, but requires real inter-individual coordination. While the technical structure of the digital ecosystem facilitates its deployment because of the modularity and granularity of projects, it must also be reinforced by the existence of social norms that promote overall coordination and minimize anti-social behavior. However, these social norms are neither those of the market (since these commons develop outside the market), nor those that would be imposed by a hierarchy (a commons does not rely on centralized regulation like a company). In fact, the coordination of individual productive activities takes place through organizational rules and routines that have gradually emerged through collective learning.

Benkler takes the example of Wikipedia. He points out that its creators did not enact a set of normative rules for the community of contributors, apart from the principle of neutrality, which requires for each article an impartial presentation of the different opinions or hypotheses on a given subject. These have evolved over time as the community interacted and communicated, as this social system became increasingly complex and contributions grew: "We cannot deny that it grew from nothing into a major global collaboration among thousands of contributors and is a system that is fundamentally collaborative and built on discussion and mutually shared norms" (Benkler 2011, p. 158). For Benkler, Elinor Ostrom's work has provided insights into how cooperative standards are easier to meet when individuals have the opportunity to participate collectively in their development[17]. This dimension is fundamental because it conditions the sustainability of all social information production in the digital ecosystem. However, the contribution of the BCIS's legal experts remains relatively modest on this issue of governance compared to the contribution of Ostrom, as we will see below.

## 1.3. The creative commons in the field of works of the mind

### 1.3.1. *Incarnation of free culture practices*

Creative communities are one of the components of information communities in the digital ecosystem, in the more restricted field of intellectual works. They all embody expressive forms that fall within the field of free culture. Lessig defines them as resources that anyone can use without asking permission. They cover resources belonging to the perimeter of free culture:

> By "the Commons" I mean a resource that anyone within a relevant community can use without seeking the permission of

17 Numerous research works have developed this aspect in the specific case of Wikipedia. Among them are the work of Cardon and Levrel (2009), Broca (2013) and Cardon (2017).

anyone else. Such permission may not be required because the resource is not subject to any legal control (it is, in other words, in the public domain). Or it may not be required because permission to use the resource has already been granted. In either case, to use or to build upon this resource requires nothing more than access to the resource itself (Lessig 2006, p. 198).

Such creative commons are not consubstantial with the digital ecosystem. Indeed, they originate from the first copyright legislation that sought to promote a balance between a space dedicated to the development of the commercialization of intellectual works (linked to their privatization) and a space dedicated to the development of free culture or creative commons[18]. It is not because the technological environment is changing that this founding principle should disappear. From the outset, the law has protected the possibility of creating commons, and the challenge today is to continue to do so by adapting the structure of cyberspace by removing its excessive power of control.

A "read only" culture, emblematic of the commercial culture developed by cultural industries throughout the 20th century, is developing, without suppressing it, a "read write" culture. These two types of culture have an unprecedented potential for development in the digital ecosystem. Lessig (2008) devotes an entire book, *Remix: Making Art and Commerce Thrive in the Hybrid Economy*, to describe the historical evolution of these two forms of culture. The "infernal machines" of the early 20th century enabled the development of a new, reproducible and commercial culture. For some, they also weakened a form of amateur culture in the sense that, from that point on, the relationship with culture placed everyone in the position of "consumer" rather than "producer". Culture would have lost its democratic scope by becoming the product of an elite, of a cultural monarchy. Lessig quotes a famous 20th century American music composer, John Sousa, whose fear was that children would become indifferent to the practice of music if it could come into the home without work or effort: "Amateurism, to this

---

18 "We can architect cyberspace to preserve a commons or not. (Jefferson thought that nature had already done the architecting, but Jefferson wrote before there was code.) We should choose to architect it with a commons. Our past had a commons that could not be designed away; that commons gave our culture great value" (Lessig 2006, p. 198).

professional, was a virtue – not because it produced great music, but because it produced a musical culture: a love for, and an application of, the music he re-created, a respect for the music he played, and hence a connection to a democratic culture" (Lessig 2008, p. 27).

Lessig is not the only one to make such an observation. In his book *Convergence Culture*, Henry Jenkins (2013) reaches similar conclusions. The commercialized forms of entertainment that emerged in the second half of the 19th century in the United States drew heavily on traditional, amateur cultural productions. They gave rise to cultural industries by producing a culture adapted to the greatest number of people. At the same time, however, they also reduced the popular cultural practices of the past to a form of "clandestinity" (Jenkins 2013, p. 170). In reality, the practice of a participatory amateur culture in the 20th century did not disappear, but it was crushed by another form of culture, a so-called mass culture, which provided access to an unparalleled quantity and diversity of culture. This new culture is also called "read only", because it introduces a form of passivity in the reception of works symbolizing cultural consumption in the 20th century[19]. For a not insignificant part, it has been transformed into a professional and commercial culture, leading to a new relationship with music, making the listener a simple consumer and no longer an active participant (in the sense of a cultural content creator).

At the dawn of the 21st century, the emergence of new digital technologies facilitated the awakening of an amateur popular culture that had lost its scope in the 20th century. Borrowing from the language of geeks, Lessig calls it the "read write" (RW) or remix culture. If in its forms of expression, this culture is new, it often corresponds to a translation in the digital ecosystem of existing sampling practices in the field of music and "collage" in the artistic field. The explosion in the number of blogs also constitutes, in the field of written media, an illustration of this RW culture.

Regardless of the expressions of this RW culture, from the point of view of traditional esthetic standards, both Jenkins and Lessig recognize that this participatory culture is often of poor quality or discloses content of very

---

19 Both Jenkins and Lessig do not in any way question the numerous studies on reception that have demonstrated the absence of passivity of the receiver. Rather, their aim is to defend the fact that the transceiver-type communication scheme confines each of the two categories of actors to a specific function: to produce or to consume.

uneven quality. But this is not the point, because the point of developing a participatory culture is to have beneficial effects on society. This is why it must be protected from any attempt to enclose it by considering it as a commons.

### 1.3.2. *Institutionalization of free culture: Creative Commons licenses*

Protecting and promoting free culture practices in the digital ecosystem requires legal recognition. However, based on the observation that politicians had tended to act in the direction of strengthening control over access to cultural resources in the digital ecosystem and that it seemed that legislators were not inclined, in the near future, to defend the existence of cultural commons, some legal experts at the Berkman Center have invented a legal solution that would nevertheless allow free culture to flourish. This solution makes it possible to create creative "public commons" through a legal subterfuge consisting of using a non-exclusive license based on private law (contracts), whereby the creator authorizes, without requesting prior permission, the copying of his work by a third party (while still acknowledging his authorship) as well as certain associated uses:

> Creative Commons has used private law to build an effective public commons. Creative Commons offers copyright holders a simple way to mark their creative work with the freedoms they intend it to carry. That mark is a license which reserves to the author some rights, while dedicating to the public rights that otherwise would have been held privately. As these licenses are nonexclusive and public, they too effectively build a commons of creative resources that anyone can build upon (Lessig 2006, p. 199).

The creators of these licenses, Lawrence Lessig, James Boyle and Hal Abelson, all three members of the BCIS[20], drew heavily on open-source software licenses. Lessig says the creation of Creative Commons, a Massachusetts-based non-profit organization based at Stanford University, "aim[s] to build a layer of reasonable copyright on top of the extremes that

---

20 Hal Abelson is also one of the founding members of the Free Software Foundation, which invented free software licenses.

now reign. It does this by making it easy for people to build upon other people's work, by making" (Lessig 2002, p. 117). Creative Commons (CC) licenses are recognizable in the form of labels associated with computer-readable metadata of cultural content.

In legal terms, they do not consist of overturning copyright legislation, but of circumventing it. They simply offer more freedoms, beyond what traditional fair use also allows. They imply that the creator agrees to lose some control over his work and, in the first place, his right to exclude anyone from copying. However, they do not imply a total loss of control. With this new system of ownership, the creator has the ability to decide himself on the degree of freedom granted by users to use his work. In this sense, these licenses offer the possibility not only to reconstruct but also to broaden cultural commons in the digital ecosystem. Cultural commons characterize both resources that can be used without prior permission (but threatened), such as those in the public domain, and new resources created in the digital ecosystem and voluntarily shared.

Associated with these licenses is the emergence of a broader conception of property where the social relationship, embodied here by sharing, takes precedence over the relationship of the individual to the object, as the economic law specialist Pierre Crétois points out:

> A right to administer access to property is, at its core, a translation of the owner's right to control his or her thing, a right seen as a social relationship rather than a mere right to preserve independence from others. The owner's power of control certainly gives the right to exclude anyone, but also to include according to the same principle: it is therefore a social relationship (Crétois 2014, p. 325, author's translation).

The foundations of this license are very close to those created by the free software movement. Lessig makes no secret of it. First of all, he borrowed the adjective "free" to describe culture, as Stallman had previously done for free software. He devotes the first chapter of his book *The Future of Ideas* (2001) to the history of this movement, to the specificity of these licenses in terms of regulation and to their economic value.

His objective is then to show that these characteristics can be extended to other types of content:

> This feature of open code, however, is not limited to code. The lesson of open code extends to other contents as well. As we will see when we consider the law of copyright, this balance between free and controlled resources is precisely the balance that the law must strike in intellectual property contexts generally (Lessig 2001, p. 72).

Free licenses, first and foremost the GPL/GNU license invented by Richard Stallmann, have thus strongly inspired the creators of Creative Commons licenses in the development of their content. In concrete terms, these licenses authorize the use of content by a third party without the latter being obliged to request authorization from its creator. In the field of software, they guarantee users four freedoms: they can freely use, copy, modify and distribute (including modified versions)[21].

In the same vein, CC licenses all allow, at a minimum, the possibility of copying cultural content without asking permission from the creator (this is the CC clause). Second, they are distinguished by the degree of permissiveness they allow. With this new system of ownership, the creator has the possibility to decide for himself the degree of freedom granted to users of his works. He can choose a license that allows any use, as long as the attribution of authorship is given. This is the license called CC BY, which is the one that gives users the greatest freedom of use. However, other CC licenses allow, on the other hand, us to restrict to a non-commercial use (NC clause), with or without modification of the original work (ND clause)[22]. Finally, to ensure that the CC-licensed cultural resource remains shared once it has been appropriated by other users, the creator can attach the SA (Share Alike) clause to it. Such a clause avoids ownership of a shared resource. Thus, this licensing system opens up a wide spectrum of possibilities that outline new creative ecologies, both commercial and non-commercial, depending on the licenses chosen by the creators. These licenses can be used as much for individual creations as for collective creations. In all cases, by their plasticity

---

21 For a detailed discussion of these licenses, see Sébastien Broca's book *Utopie du logiciel libre* (2013).

22 For a detailed explanation of all Creative Commons licenses, visit the official website at https://creativecommons.org/.

and variety, these CC licenses have the function of accompanying free creative practices in the digital ecosystem and the (re)construction of new cultural commons.

Let us take the example of the Wikipedia encyclopedia. All articles, "multi-handed" products and open access systems are protected by a Creative Commons type property system. Originally, Jimmy Wales, its founder, had chosen to protect published articles with the GPL license, originally used for free software. It was only in 2009 that the Wikimedia Foundation, which manages Wikipedia, adopted the CC BY-SA (authorship – share alike) license for its content. This is a *sine qua non* condition for any contributor who wishes to publish a new article. This choice is consistent with the objective of collaborative writing on each article, each contributor being invited to intervene in discussion spaces to enrich in turn the articles published according to a continuous process. Sharing with modification is therefore mainly aimed at the community of contributors. The SA clause means that any user, contributor or not, can access the content, as well as republish it in its entirety on the condition that he leaves it himself under the same type of license. This clause conditions the development of a common cultural heritage. It should be noted, however, that Lessig does not specify whether only creative practices using the SA clause fall within the scope of the commons. In this sense, he pays more attention to the question of the construction of the commons than to their sustainability and the modalities of their governance. Now, this is a question that can legitimately be asked, because if the stake is to protect a whole set of resources from possible threats of privatization, this clause constitutes a very effective defense. This point is now being debated in the circle of defenders of the commons, who do not all share the same position.

### 1.3.3. *The modalities of cohabitation with the commercial cultural economy*

The creative commons economy is not easy to grasp, because while it brings together creative and sharing activities in the non-market economy, it has also spread beyond this boundary in what Lessig calls a hybrid economy. In both cases, the main actors are the voluntary contributors who create cultural content (text, sound, image), essentially amateurs, and the hosting and distribution platforms for this content, which institute a new form of re-intermediation. The central intermediation function devolved to producers and publishers in the cultural industries is fading or even disappearing altogether to give way to direct relations between creators and the actors of sharing and distribution. In the digital ecosystem, the cultural economy of the commons is rather a combination of several arrangements that maintain singular links with the commercial cultural economy.

### 1.3.3.1. *The non-market arrangements of creative commons: unfair competition?*

The non-market arrangements of the commons include all the practices of creation, sharing and distribution of free cultural content, whether individual or collective, which do not give rise to economic transactions in monetary form, whether at the level of creators or sharing platforms. These different activities are not without effect on the commercial cultural economy, as they introduce, without intending to do so, a form of indirect competition that destabilizes the production routines of the traditional players in the cultural industries markets. For BCIS jurists, these practices do not constitute a form of unfair competition, even if they have often been perceived as such. They inevitably introduce a questioning of the economic models that until now have structured these industries without, however, implying their disappearance. A cohabitation is possible, but it is still necessary that the latter be able to develop their strategies to, at worst, survive or, at best, take advantage of the existence of these new creative arrangements. Let us give a few examples, each of which sheds light on this question.

The very significant growth in the early 2000s of networks for exchanging cultural content files via peer-to-peer platforms, which was mostly part of the non-market economy (the platforms did not derive direct profits from their activity), was considered a major threat to the music industry in particular, which was the first to be affected by these practices. Faced with the rise of creative forms of sharing and transformative practices that could harm their trade, the actors of these cultural industries then lobbied the legislator for a new legal framework to eliminate these forms of competition that they considered unfair.

For BCIS jurists, rather than condemning these practices as illegal, it is more judicious to offer them a new institutional framework that would lead to the deregulation of amateur creativity and thus facilitate a harmonious cohabitation. In particular, they argue that rights owners could receive compensation for potential losses related to this legalization in return for the legal recognition of amateur practices of sharing copyrighted works. Both Benkler and Lessig refer to academic William Fisher's[23] proposal for a principle of compensation in the form of a global license financed by a tax or by the federal budget[24] to legalize and finance the sharing of cultural works in non-market arrangements. While they reject the idea that this principle could replace the copyright regime, they admit that it could facilitate the functioning of a more competitive market: "This competitive regime, with a back screen to ensure that artists do not lose, would facilitate a great deal of innovation in the distribution of content" (Lessig 2009, p. 120).

On the other hand, if a commercial platform uses the products resulting from this amateur creativity for commercial purposes, one leaves the strictly non-market framework and must then remunerate the authors directly. Lessig gives a very clear example: "If a parent has remixed photos of their child with a song by Gilberto Gil (as I have done on numerous occasions), then when YouTube makes the amateur remix available to the public, Gil must be compensated in some way" (Lessig 2008, p. 256). The problem is that the law considers amateur sharing illegal but, at the same time, provides a form of immunity to sharing platforms that are not responsible for the hosted content and therefore have no form of obligation to compensate rights owners arising from making it available for undifferentiated uses[25].

The emergence of Wikipedia, emblematic of a mode of social production of cultural commons taking shape in a non-market setting, has also gradually contributed to shaking up the routines of traditional actors. Wikipedia, it should be remembered, is produced by thousands of volunteer contributors who cooperate to producing a cultural commons. The fact that each article is protected by a Creative Commons license is the main marker of a sharing

---

23 William Fisher is a Harvard professor and a specialist in intellectual property. He is also the current director of the Berkman Center for Internet and Society. He wrote a book in 2004, *Promises to Keep: Technology, Law and the Future of Entertainment* (Stanford University Press).

24 As will be shown in Chapter 2, the idea of a global license has been the subject of much attention and debate but has not, with rare exceptions, attracted the support of cultural industry representatives.

25 This point will be discussed in more detail later because today, the latest European legislation on copyright revision dating from 2019 has endorsed the principle of the need for commercial platforms to enter into agreements with rights owners (in return for remuneration for the platform) so that their users can use their content without being threatened with prosecution.

economy, with each user being able to appropriate the content without asking permission. The contents of the encyclopedia are also distributed by a technical platform that is financed outside the commercial sphere by donations. It is undeniable that the rise of Wikipedia and its appropriation by a growing number of users has contributed to strongly destabilizing the traditional markets for encyclopedias, which had already had to adapt their economic model to the digital ecosystem (Shapiro and Varian 1999). Here, too, we are in a configuration where economic players, such as the *Encarta* encyclopedia proposed by Microsoft or the oldest English encyclopedia, *Britannica*, find themselves confronted with a singular situation: "Both companies found themselves in competition with a business model that simply did not exist a decade ago; a model so implausible that is theoretically could not exist, or so we thought until just a few years ago" (Benkler 2011, p. 212). The founders of Wikipedia neither wanted nor imagined they could destabilize the encyclopedia market. For some time, moreover, it was the object of much criticism regarding the low quality of the articles offered compared to the expertise of traditional encyclopedias. However, the progressive audience of this open encyclopedia has gradually upset the players in place, who have been forced to change their strategy and, in turn, to be creative in order to hope to remain on the market.

Finally, the commercial cultural economy can even benefit from these non-market agencies of the commons. For this to happen, these commercial actors must agree to reduce their control over the cultural content they produce and recognize that this strategy can be a source of economic value by offering them the opportunity to expand the boundaries of their market. For example, a creator can, in fact, see the value of his cultural property increase if he decides to make it exist in both market and non-market economies. Lessig cites the example of Cory Doctory, a science fiction author, who distributed his latest novel for free on his blog under a Creative Commons license the same day it was offered for purchase in bookstores. By offering potential consumers the opportunity to appropriate it in digital format, it increases its value because the book is an experience good. However, in this case, it requires adapting the contracts between creators and publishers who, most of the time, do not authorize this type of hybrid practice. In the case of music, Sacem in France prohibits an artist from adopting such practices.

Producers can also reap substantial benefits at another level. Lessig cites the example of ccMixter (in the United States), a non-profit community music site promoting remix culture, which offers CC-licensed content and is funded by donations from contributors. This type of sharing platform can also build a new kind of relationship with producers who can find new artists to produce while minimizing the transaction costs (mostly research) associated with finding new talent: "When labels discovered artists in ccMixter and then signed them to record deals or contracts, the work the artist had freely licensed continued to be free. Indeed, sometimes the very same song was licensed both commercially and noncommercially. This helped the commercial. More artists and record companies will do the same in the future" (Lessig 2008, p. 226).

### 1.3.3.2. *Hybrid arrangements of creative commons*

There is a part of the cultural economy that produces commons that is based on a so-called hybrid mode resulting from the production of voluntary contributors who will create content, individually or collaboratively. By using CC licenses, contributors signal that they belong to a sharing economy. However, the platform that makes cultural content accessible will monetize this content for its own benefit (and not on donation or public funding). From this perspective, there is a strong porosity with the world of the commercial cultural economy without being able to be fully assimilated into it. Indeed, these hybrid arrangements are distinguished by the willingness of the actors to preserve the separation of these two spheres, market and non-market.

However, this boundary is unstable and threatens to collapse if no attention is paid to the specific conditions under which it is maintained. In particular, contributors may no longer be willing to produce cultural content on a voluntary basis if they feel that they are merely free inputs to a commercial economy. In other words, the agents of the commercial economy (the platforms) that exploit the value created by voluntary contributions must ensure that they maintain an ecosystem that promotes sharing by proposing operating rules that encourage contributors to create. They must seek to understand and internalize the norms of the community they are exploiting: "A key element to a successful hybrid is understanding the community and its norms. And the most successful in this class will be those that best leverage those norms by translating fidelity to the norms into hard work" (Lessig 2008, p. 184). It is at this level that Lessig's approach reflects the importance he places on community. If the standards that structure them are not internalized by commercial platforms, then there is a risk of dilution of the commons due to the gradual disappearance of contributors.

Benkler agrees with Lessig's point of view, stating that one of the challenges in maintaining these collaborative modes lies in the company's ability to organize a benevolent and non-hostile relational mode, since users cannot be directed like traditional workers. He referred to the world of Open Source, taking the example of companies like Red Hat or IBM, which have been able to develop an original symbiotic relationship with communities of Open Source software developers and make significant economic profit from it. However, he points to the same condition as Lessig: "If a company exploits its community by failing to contribute its fair share or to respect the community dynamics, it will ultimately alienate the community and the system will fall apart" (Benkler 2011, p. 219). Maintaining a mutual relationship of trust is an absolute condition for the survival of a business that uses free inputs from a community of volunteers.

Both Lessig and Benkler are betting on a reinforced cohabitation in the medium term if market organizations adapt and internalize these new conditions of production and distribution of knowledge and culture. If the world of free software is the emblem of this hybrid economy, the cultural sphere is also seeing the deployment of such economies. Among the examples given, two of them seem particularly interesting to us because they clearly reveal the stakes involved in cohabiting with the commercial cultural economy.

First and foremost, they are commercial platforms that economically exploit the value created by the contributions of a community of volunteers. In 2008, at the time when Lessig raised this issue in his book *Remix*, the photo-sharing platform Flickr seemed to him to be an emblematic case of this hybrid economy with strong growth potential. Conversely, the failure of the Ofoto platform created by the company Kodak attests to the difficulty of articulating a community logic and a 100% commercial model. The goal of Flickr was above all to build a community by facilitating the sharing of photos between its members. From the outset, 80% of the photos published were in sharing mode thanks to the use of Creative Commons licenses. Unlike Ofoto, Flickr did not have control over the photos hosted on its platform. A feeling of community belonging (more or less strong) was a central element in the choice of this platform by the contributors. When Flickr was bought by Yahoo! in 2005, the goal was not to turn the platform into a commercial site and make a maximum profit. While Yahoo! could have extracted considerable value in the form of advertising revenue due to the platform's very large audience, the company chose a less profitable business model in the form of subscriptions to some of its contributors (in exchange for significant storage space) that allowed it to remain in this hybrid[26] arrangement.

The challenge is to bring this feeling of reciprocity to life, based on mutual benefit for each of the stakeholders. Amateur photo creators have a public space for visibility and sharing in exchange for their free sharing of the content they create: "Every company building a hybrid will face exactly the same challenge: how to frame its work, and the profit it expects, in a way that doesn't frighten away the community. 'Mutual free riding' will be the mantra, at least if the value to both sides can be made clearer" (Lessig 2008, p. 237).

The question of "rewarding" voluntary contributors is not, however, excluded from consideration, but, according to Lessig, applying a principle of retribution is not an effective solution because the ethics of contribution and sharing are not the same as those of business ethics. Removing this distinction amounts to jeopardizing the hybrid economy, which would gradually dissolve within the commercial economy. We will see later that this condition of non-remuneration is not so obvious. We can indeed imagine that, just as the platform monetizes its content without making the search for profit its top priority, contributors can also have the possibility of being remunerated without necessarily harming their initial motivations for sharing. Why indeed could such a principle be considered valid for one of the players and not for both?

A final example shows the difficulty for commercial cultural actors to understand how they can benefit from an alliance with hybrid arrangements. Many "fandom" platforms have been created on the Web, such as the

26 The Flickr platform has undergone major changes since this period. Recently, it has undergone a major transformation since its managers wanted to remove all content produced before a certain period. A negotiation took place with the Creative Commons foundation so that photos shared with such a free license could be kept.

famous *Daily Prophet* newspaper, created by a 13-year-old girl, Heather Lawver. It brought together publications from young teenagers from around the world who wrote stories based on the *Harry Potter* saga. While the author, Joanne Rowling, welcomed these new spaces of fanfiction, her producer, Warner, who wanted to control all forms of products derived from the work, did not agree. This led to what has been called "The Potter War", which Henry Jenkins recounts in detail in his book on the culture of convergence. After a phase of negotiations, an agreement was reached. Warner finally agreed not to sue this amateur newspaper. After a long learning process, the production company understood how it could interact with and even benefit from fan communities. Fans were a significant part of the marketing budget that Warner did not have to pay for: "Warner had learned that being less restrictive with its intellectual property strengthened fans' loyalty to the brand and, hence, the return to its artists" (Lessig 2008, p. 211). It would be interesting to extend this reflection by evaluating how, today, the world of geek culture cohabits with the commercial world[27].

Ultimately, the creative commons economy is deployed through multiple agencies, none of which is intended to replace the traditional cultural economy. The harmonious coexistence between these different spheres is, however, conditioned by the learning of new regulatory rules by commercial cultural actors (cultural industries such as intermediation platforms) who must now coexist with amateurs in a totally new way, as the latter are no longer confined to their exclusive function as consumers.

New implicit ethical norms must indeed regulate these new cultural spaces and, in particular, confirm the feeling that everyone is in a mutually advantageous situation. The contributor must have the feeling of being "compensated" in one way or another for his or her voluntary "work", which is then exploited by a third party. This compensation need not necessarily be monetary. Likewise, the search for profit should not be the primary objective of the economic exploitation process of the platforms hosting the shared cultural content. For, in this case, the feeling of exploitation by volunteers

---

27 We can cite several works in information and communication sciences that question the development of the phenomenon of fan and geek cultures from the point of view of creative activity and production of works (fanfictions, *mashups,* remixes). See the synthesis written by Mélanie Bourdaa: Bourdaa, M. (2015). Les fan studies en question: perspectives et enjeux. *Revue française des Sciences de l'information et de la communication,* 7.

may jeopardize these hybrid economic models. Without saying so explicitly, this question refers to the institutional form of governance of these platforms. The cultural content platforms of the hybrid economy cannot be companies like any other. This raises the question of their legal status and their modes of financing. Are all economic models compatible with this hybrid economy?

The writings of the American jurists at the BCIS have had the merit of questioning the foundations of the cultural economy in the digital ecosystem. They have opened up a range of questions that have given rise to a genuine research program on the question of the cultural commons' economy (more broadly, intangible commons) as a new space for the creation, production and distribution of cultural content in a digital environment. Their contributions are multiple. They have stressed the importance of creating the conditions for an institutional ecology favorable to the development of such an economy, requiring the intervention of the legislator to amend the copyright law in order to reduce control over cultural content by showing that this can be a new lever to foster new forms of creativity and innovation. Although inevitably destabilizing the behavioral routines of the economic actors dominating cultural markets, the different facets of the cultural commons economy do not constitute a direct threat.

## 1.4. Propagation in the intellectual and militant sphere in France

In France, since the beginning of the 2000s, intellectuals of diverse origins, all militant within associations or collectives oriented towards the defense of freedom on the Internet, have contributed to propagating in the public space some of the strong ideas of BCIS jurists. This militant "commonsphere" invested in the public space on the occasion of the debates that accompanied two major bills on the adaptation of copyright to the digital ecosystem. We will evoke these two periods in France where the pioneering approach of the BCIS jurists was mobilized to defend, on the one hand, the legalization of non-commercial sharing with the proposal for the establishment of a global license (in the framework of the DADVSI law (2005)) and, on the other hand, the institutional recognition of information commons (in the framework of the law for a digital Republic (2016)).

### 1.4.1. *The challenge of legalizing non-market sharing*

#### 1.4.1.1. *Contextualization of the DADVSI law*

A first intellectual battle took place in the French public space during the process of implementation and examination of the DADVSI (*droit d'auteur et droits voisins dans la société de l'information* – copyright and related rights in the information society) (2005) and HADOPI laws ("Creation and Internet" law promoting the dissemination and protection of creation on the Internet) (2009). These two laws from the beginning of the millennium in France had a twofold stated objective: to adapt copyright to the digital age and to promote the financing of creation. It all began in 2002 when the government entrusted the CSPLA (*Conseil supérieur de la propriété littéraire et artistique*) with a mission to transpose European Directive 2001/29 on the harmonization of certain aspects of copyright and related rights in the information society.

Following this preparatory work, a draft law was adopted by the Conseil des ministres on November 12, 2003. It provides that technological measures may prohibit purely and simply all private copying in the context of works distributed by an on-demand service, or if the user who wants to make the copy has not lawfully acquired the work. Technological measures may also limit the number of private copies of an original work not distributed by an on-demand service (CD, DVD) to a single copy. The DADVSI law also warns of three years in prison and a fine of 300,000 euros for anyone who proposes, uses or makes known, directly or indirectly, a tool or piece of information, to neutralize a technical measure, regardless of the purpose pursued by the user. Such acts are assimilated in the law to counterfeiting offences, offences which are accompanied by a presumption of guilt.

The core of the project has therefore focused on the legalization of additional legal protection for rights owners in the form of DRM (Digital Right Management) as a means of combating all forms of downloading of protected works on peer-to-peer platforms. Exceptions were considered as possible options by the directive. In the end, the government law retained only two of them: the exception for disabled people authorizing them to translate works without requesting prior authorization and the legal deposit of web pages with the BnF (Bibliothèque nationale de France) and INA.

### 1.4.1.2. *The space for debate and controversy*

This DRM legalization project has sparked many debates and controversies in the public space, initiated by actors from different horizons: the Libriste movement (in particular, the French section of the Free Software Foundation with the association April), certain actors from the cultural industries grouped within a militant collective, as well as intellectuals such as Philippe Aigrain and Valérie Peugeot, who are militant for the protection of information commons on the Internet within associations such as Vecam and Quadrature du Net.

As early as 2002, the Libriste movement had set up a powerful citizen lobby through a website (Eucd.info) that proposed critical analyses of the law in order to alert the public to the harmful social and economic consequences of the European Directive and the DADVSI law. Their main criticism concerned the fact that cultural uses (reading or listening to digitized works) and its exceptions (notably the exception for private copying) would henceforth be transformed into contractualizable rights that could be arbitrarily limited by technology. On a practical level, they demonstrated strong citizen activism by contacting numerous members of parliament and putting a petition online that obtained more than 100,000 individual and collective signatures. The echo of their actions went beyond the strict framework of the software world.

For Sébastien Broca, a specialist in information and communication sciences (ICS) of open-source software:

> Librarians succeeded in starting a public debate about the threats to individual liberties linked to "internet policing" and about the new culture economy that was born from the irruption of digital exchanges. They brocaded the attitude of the music and film majors, accused of wanting to maintain at all costs their economic model, based on scarcity and control of usage (Broca 2013, p. 179, author's translation).

On the side of the representatives of the cultural industries, while most producers, audiovisual and multimedia broadcasters and publishers defended this project, authors, artists and performers did not form a homogeneous group like the collective management societies. In particular, a collective under the name Alliance Public-Artistes, bringing together Adami and

Spedidam[28] as well as the association UFC-Que choisir and the Union nationale des associations familiales, undertook collective action for recognizing the legalization of non-market sharing on peer-to-peer platforms. In 2005, they created a dedicated site to present their project, defending a compensation system in the form of a global license: "The global license is an authorization given to Internet users to access cultural content (music, images, films, texts) on the Internet and to exchange it between them for non-commercial purposes in exchange for a remuneration paid to artists when they pay their monthly Internet subscription".[29] We find here the idea previously defended by Lessig and Benkler. Introducing this new principle of financing creation implied *de facto* the legalization of acts of remote downloading by recognizing that they fall under the exception for private copy, on condition that they remain in a strictly non-commercial framework. Internet access providers would be the entities responsible for collecting, via their subscription, the sum intended to compensate for these acts of private copying.

The Alliance Public-Artistes collective had committed itself in favor of a so-called optional global license, not generalizable to all Internet users. Each one, at the time of his subscription, should promise not to engage in acts of downloading likely to be exempted from a supplement. On many occasions, the members of the collective have regularly intervened in the space of parliamentary debates. They have also requested a report validating the legal feasibility of a global license. They mention on their website an Ipsos poll, which shows that three quarters of French people are in favor of such a compensation principle.

This idea of a global license, initially defended by BCIS jurists, had also found a favorable echo among militant associative actors such as Philippe Aigrain and Valérie Peugeot. Together with the European deputy and former minister Michel Rocard, Jacques Robin, founder of Transversales Sciences Culture, and Patrick Viveret, member of Transversales Sciences Culture, they co-wrote an article in the newspaper *Libération* on July 29, 2004, expressing their opposition to the DADVSI law[30].

---

28 Adami and Spedidam are civil societies for the administration of performers' rights.
29 www.lalliance.org/pages/2_1.html.
30 A summary is available at: https://vecam.org/archives/article335.html.

Philippe Aigrain, an intellectual figure with an eclectic background, in turn researcher–computer scientist, entrepreneur, employee of the European Commission, expert on free software issues and essayist, is the one who contributed to making the Berkman Center's approach to free culture and creative commons known in French intellectual circles in the early 2000s. He had become aware of the theories of the Berkman Center's jurists very early on. In 2003, he wrote a column in the newspaper *Libération,* entitled "Pour une coalition des biens communs" (For a Coalition of Common Goods), in which he advocated the formation of a political alliance in favor of the protection and promotion of all forms of cooperative production of high social utility information commons, threatened by an information capitalism that privatizes all forms of knowledge, headed by multinationals with ever-increasing financial power. The same year, he edited a special issue of the *European Journal for the Informatics Professional* with the Spanish researcher Jesus Gonzales Barahona, on the theme of Open Knowledge. Yochai Benkler was invited to participate in this issue and presented a summative version of his seminal article "The Political Economy of Commons". Philippe Aigrain subsequently published two books, *Cause Commune* (2005) and *Sharing Economy* (2012), in which he sets out his vision of commons and the solutions to be implemented to protect them and promote their development with the introduction of a creative contribution.

### 1.4.1.3. *From global license to creative contribution*

Like the global license principle, this idea of creative contribution is based on a mandatory lump-sum mutualized financing and establishes a disconnection between uses and payment. However, it differs from it in terms of purposes, management and distribution modalities. The creative contribution is not a principle of compensation for right owners, but rather a new form of remuneration not so much exclusively for the latter as for all creators, amateurs in particular, who deliberately choose to make the product of their cultural creative activity freely available. It is conceived as a new form of financing for digital creative culture promoting the creation of knowledge commons.

For its creator, Philippe Aigrain, the primary challenge is to promote quality cultural production in all media, which is sustainable in the medium term in a world where everyone's cultural commitment is increasingly important. Many contributors to cultural commons do not seek direct monetary remuneration, because they are sensitive to other forms of symbolic remuneration and because they devote part of their lives to "leisure" and not to their work activity as such (the one that gives them an income). However, such so-called amateur production can only develop and participate in cultural diversity if it finds levers of remuneration so that everyone can devote more time to it, a guarantee of essential learning in order to hope to increase the overall level of quality. In terms of management modalities, as this license is not anchored in intellectual property law, it is not intended to be managed by collective management societies, as envisaged for the global license, but rather by an independent and transparent body. Finally, while the global license is based on a strict distribution of right owners according to the assessed levels of downloads, the distribution of the creative contribution is based on actual usage and establishes a principle of correction, so as to protect cultural diversity. A fraction could even be used to finance cultural projects or organizations.

Such a creative contribution license would also prevent the economic value created by amateur cultural production left freely accessible from being captured and reappropriated by actors who know how to monetize this content by making the wheels of the attention economy turn to their advantage, as Lionel Maurel[31] points out in support of this proposal. A final interesting argument is also put forward. According to him, this legalization would make it possible to promote strictly decentralized exchanges that could balance the growing trend towards centralization (stemming from the activity of distribution platforms).

### 1.4.1.4. *The outcome of controversies*

Let us now return to the outcome of the DADVSI law. Even though it was opposed to any form of legalization of sharing, even non-commercial, the debate on the public space, then relayed in the parliamentary space around this proposal for a global license, had a surprising effect since two identical legislative amendments in favor of this proposal (each carried by a different political party, UMP and PS) were tabled and received, to the surprise of all, a majority vote a few votes apart. But this parliamentary *coup de théâtre* ended with a final episode in March 2006, when the parliamentary session reopened, which put an end to this amendment legalizing non-market sharing

---

31 All these arguments are developed on an article in his blog at the following address: https://scinfolex.com/2012/11/06/reponse-aux-arguments-du-parti-pirate-suedois-contre-la-licence-globale.

by voting a contrary amendment. It was the repressive aspect that was finally ratified with the possibility of prosecuting the authors of file exchange software likely to allow the illegal exchange of works (which undermined the principle of technical neutrality) and the principle of a three-year prison sentence and a 300,000 euro fine for any Internet user who illegally made works available with peer-to-peer software.

While this proposal for a global license has only had ephemeral institutional recognition, it must be said that the space for debate has not ended with its early disappearance. It returned to the forefront during the Hadopi law in 2009, without more success, only to be definitively rejected a few years later in the Lescure report (2013), which was to serve as a basis for a new bill on cultural exception II[32]. In this respect, a very didactic map of the controversies surrounding this question of global licensing has been proposed by the medialab of Sciences Po Paris[33]. However, this report did not deny the interest of legalizing non-market sharing in the face of the crisis of confidence that had developed between the cultural industries and some of their public:

> The legalization of non-market exchanges would promote access for everyone to all cultural content available online and would valorize the notion of disinterested sharing... The test introduced in return for the legalization of exchanges, whether in the form of compensatory remuneration or a 'creative contribution', would provide creators with a substantial source of income (Lescure 2013, p. 31, author's translation).

This report refers not only to the principle of global licensing but also to the principle of a creative contribution proposed by Philippe Aigrain (2012) in his book *Sharing, Culture and the Economy in the Internet Age*. Solving this crisis was even the major axis of this report. In this perspective, it invited taking into account the difference between occasional downloading practices for purely private purposes and without the objective of enrichment and the lucrative activities deployed by certain Internet actors who systematically exploit the distribution of counterfeit cultural goods.

---

32 For more details, refer to the cartography of the controversy published by the médialab Sciences Po Paris: http://controverses.sciences-po.fr/archive/licenceglobale/etape-6/index.html.

33 https://controverses.sciences-po.fr/archive/licenceglobale/.

Despite this observation, the report concludes that not only does the legalization of non-market trade raise legal obstacles[34] that would amount to calling into question the 2001 European Directive, but that it also faces economic and socio-technical obstacles. In particular, the coexistence of legalized non-market exchanges and a commercial offer seems difficult to envisage, at least in the short term, because if all content became free instantly, then paying platforms would have difficulty attracting customers. Another argument put forward is the excessive financial amount of this proposal for a creative contribution for certain households if it were to concern all content protected *a priori* by copyright. Finally, the report is concerned about the technical measures to be implemented, which would consist of systematically observing traffic, which could be detrimental to the respect of individual liberties. Thus, he argues rather in favor of the sustained development of subscription offers (in the form of music streaming, subscription video-on-demand, etc.) which, compared to the introduction of private global licenses, "seems more respectful of the freedom of rights owners but also of users, who can thus choose the type of content as well as the type of services they wish to have access to" (Lescure 2013, p. 355, author's translation). In the medium term, however, it is accepted that financing through a creative contribution of a "free" offer may be conceivable....

### 1.4.2. *The challenge of legal recognition of the information commons*

A few years later, another intellectual battle led by commonsphere activists in France focused on the issue of positive recognition of the information commons. Initially, this proposal appeared during an online consultation opened by the *Conseil national du numérique*, whose Vice-President at the time was Valérie Peugeot. She was not included as such in their subsequent report

---

34 In order to make such a proposal legal, it was necessary either to recognize the legality of the principle of exhaustion of rights in the digital ecosystem, stipulating that the exclusive rights of the owner disappear on the first sale or first circulation of the medium containing the protected work, provided that this circulation was carried out by the owner of the rights or with his consent, or to introduce a new exception to copyright for private copying. However, the report concludes that both of these eventualities run counter to European Directive 2001/29/EC. Indeed, the latter stipulates that copyright and related rights are not exhausted in the digital ecosystem, nor does it provide for this possibility for private copying in the list of optional exceptions. The revision of this Directive is the only possible way forward, and therefore, in other words, a solution that is unlikely to be envisaged in the short term.

"*Ambition numérique*" (digital ambition), although the report[35] states that commons based on collective action and a mode of production and of common governance constitute both a new political space and a new relationship to value, which is essential to support. It reappeared on the scene during the consultation phase on the Internet in the context of the bill for a Digital Republic[36] (*République numérique*). This project, led by Secretary of State Axelle Lemaire, aimed to develop a legal framework to build a true "data economy". As mentioned by the common property activist Lionel Maurel in his detailed account of the genesis of this proposal[37], Axelle Lemaire wanted a proposed article to be submitted for discussion on the information commons. The choice was then made to convene article 714 of the Civil Code, which deals with "common things" in order not to confine the commons to the field of intellectual property. Their positive recognition would allow an easier recourse to the judge to counter all attested copyfraud practices. This is how the first proposal of Article 8 came about, which was filed on the *République numérique* platform:

> The following fall within the common domain of information: 1) information, facts, ideas, principles, methods, discoveries, data, as long as they are subject to public disclosure in compliance with the laws and regulations in force and are not protected by a specific right; 2) objects protected by an intellectual property right, or by another exclusive right, whose legal protection period has expired; 3) information from administrative documents that are publicly disseminated.

In this formulation, it can be seen that it comes close to the definition of the structural public domain as defined by the association Communia. On the other hand, it excludes from its perimeter the so-called voluntary commons. In its initial form, this proposal had only a very limited lifespan. It was quickly modified following ministerial discussions, which, according to Lionel Maurel,

35 Conseil National du Numérique (2015). Ambition numérique. Pour une politique française et européenne de la transition numérique [Online]. Report, June 2015. Available at: https://contribuez.cnnumerique.fr/sites/default/files/media/CNNum--rapport-ambition-numerique.pdf.

36 Legifrance (2016). Loi no. 2016-1321 du 7 octobre 2016 pour une République numérique [Online]. Available at: www.legifrance. gouv.fr/affichLoiPubliee.do?idDocument=JORFDOLE 000031589829&type=general&legislature=14.

37 Maurel, L. (2017). La reconnaissance du "domaine commun informationnel": tirer les enseignements d'un échec législatif [Online]. Available at: https://hal.archives-ouvertes.fr/hal-01877448/document.

removed what constituted the essential elements of this founding proposal. Subsequently, a collective under the name of Soutons les communs, bringing together 13 associations, some of which were previously mentioned such as Quadrature du Net, SavoirsCom1 and Vecam, made proposals for amendments to this article. Box 1.1 presents a summary[38] of these proposals.

---

1) Introduce criminal sanctions for infringements of the ICD (Information Common Domain) in the same way that there are sanctions for copyright infringement.

2) Include faithful reproductions of two-dimensional works in the definition of the ICD, because too often the acts of digitization of public domain works constitute a pretext for claiming new rights that hinder the reuse of reproductions.

3) Avoid the legalization of copyfraud, because as it stands, Article 8 paradoxically risks leading to the legalization of abusive reappropriation practices: if the information, facts and ideas cannot be the direct object of a property right, on the other hand, if they are included in a database, the database can be the object of a property right.

4) Recognize a legislative existence for voluntary commons, that is, resources voluntarily shared by their creators (in particular works under Creative Commons) in order to protect them against any attempt at abusive reappropriation.

5) Create a national public domain registry to identify intellectual works in the public domain through the BnF (Bibliothèque nationale de France), which, through its catalogs, has a large amount of metadata for calculating the duration of rights.

---

**Box 1.1.** *Proposed amendments to Article 8 of the Law on the Digital Republic (République numérique)*

Following these proposals, Internet users were called upon to react. Article 8 was the one that received the most comments and 80% of favorable opinions. However, a strong coalition of representatives of rights owners (SEPM, SACD, SNEP, SNE) strongly opposed it and lobbied hard to have Article 8 removed from the bill, claiming that it presented the public domain as the rule, with intellectual property being relegated to the realm of the exception. French law was, in their view, effective in combating abusive intellectual property claims. In the end, the fate of Article 8 was sealed by an

---

38 http://soutenonslesbienscommuns.org/contributions/#SAVOIRS_COM_1.

arbitration rendered by Matignon, who decided to withdraw it before the bill was introduced in parliament. Institutional lobbying was victorious over citizen lobbying, at least in this intellectual battle.

Along the way, some of the representatives of the commonsphere recognized that this Act nevertheless advanced the commons in the field of knowledge. Valérie Peugeot expressed her views on this subject a year later on Vecam's website. First of all, she asserted that the inclusion "of the word 'commons' in law is neither an operational necessity nor an imperative for the commons to survive and develop. On the other hand, we formulate the hypothesis that this statement is necessary so that the diversity of the practices of the commons, which make up their richness but also their weakness, can set in motion a dynamic of collective construction of a unifying political horizon"[39]. With regard to the law for a digital republic, it also recognizes a real advance concerning the commons in the field of scientific research. Indeed, this law formalizes the institutional recognition of the principle of open access for public research (Article 17), allowing researchers to freely disseminate their scientific works, mostly financed by public funds, while respecting a period of exclusivity for the benefit of the publisher of 6 months for science and technology and 12 months for the humanities and social sciences. It also authorizes "text and data mining", that is, the right for researchers to use tools for massive and automatic searches of document corpuses, which constitutes a second encouraging exception to copyright. Although there are regrets that this law may have a limited scope by not making free publication mandatory, which is left to the free choice of the author, it recognizes that, henceforth, "knowledge can be explored by digital methodologies to feed the next generation of research, thus dissociating the intellectual property rights of scientific publishers from the usage rights of researchers"[40].

Undeniably, these multiple intellectual battles have contributed to bringing to the forefront the socio-economic stakes, in terms of innovation and creativity, related to the protection, recognition and enhancement of common knowledge and culture. By way of illustration, we would like to give two examples.

---

39 Peugeot, V. (2016). Facilitatrice, protectrice, instituante, contributrice: la loi et les communs [Online]. Vecam, September 29, 2016. Available at: http://vecam.org/Facilitatrice-protectrice-instituante-contributrice-la-loi-et-les.

40 *Ibid.*

In 2012, Sacem and the Creative Commons France association signed an agreement authorizing members of this collective management organization to place their works under one of the three Creative Commons licenses that allow the distribution of works for non-commercial purposes. Since January 1, 2012, Sacem has registered 1,356 works under Creative Commons licenses, which have been added to its repertoire by 138 of its members (authors, composers, directors and publishers, mainly men, who own a total of 12,461 works). However, this still represents a very small portion of its rights owners. The non-commercial restriction has been the subject of some strong criticism from defenders of free culture, such as the association Musique Libre, which publishes the Dogmazic platform[41]. Lionel Maurel is also concerned about the vague interpretation that can be given to the non-commercial use clause, as it is likely to hinder collective uses. Libraries, for example, cannot use pieces of music under NC license to sound their spaces without paying royalties to Sacem, within the framework of the general performance contract that normally binds them to the collective management society. In spite of this, this agreement is the beginning of recognition for an institution that has always expressed a strong reluctance towards the emergence of free licenses.

The second example that we thought it interesting to cite is the report commissioned from Joëlle Farchy in 2017 by the CSPLA[42] on an inventory of the state of play on the use of open-source licenses in the cultural field. Generally speaking, the recommendations proposed are in the direction of a better recognition of creative practices using this type of license. Among them, we can note the importance given to a better communication of these licenses to the public, the creation of a collective management organization dedicated to creators who opt for free licenses, or the possibility of making projects under free licenses eligible for public funding (such as CNC or CNL). The succinct conclusion given to this report shows a step forward in the recognition of free culture by institutional actors such as the CSPLA:

> All of these concrete proposals aim to include free licenses, which are contractual tools that are sometimes still controversial, in the structure of common projects and specific

44 www.numerama.com/magazine/21469-l-accord-sacem-creative-commons-sous-le-feu-des-critiques.html.

42 Farchy, J. (2017). Les licences libres dans le secteur culturel. Mission report for the CSPLA, December.

cultural productions. Behind the rough appearance of legal-technical tools, the stakes are high: the emergence of collaborative or transformative works, the development of participatory artistic projects, dissemination to a wider public, new opportunities for value creation, or even improved accessibility, outreach and enhancement of public institutions' resources (Farchy 2017, p. 58, author's translation).

We cannot end this section without mentioning the new European Copyright Directive that was ratified in March 2019. It is clear that the issue of free culture and cultural commons is no longer at the center of the debate. It has moved to the level of the strong tensions existing between the cultural industries and the digital giants. However, the voice of commonsphere activists has also made itself heard to defend, after some hesitation, this directive and, in particular, Article 13 (now 15), which requires platforms to enter into agreements with rights owners so that they are remunerated when a user posts on the platform a work (a text, a song, a film, etc.) for which they hold the rights. This obligation profoundly transforms the status of the GAFA. La Quadrature du Net and Lionel Maurel[43] defended the idea that this law was not a defeat for a free and open Internet and for the dissemination of knowledge commons. From a status of passive hosts benefiting from an attenuated responsibility towards the acts committed by their users, commercial platforms will now have to assume responsibility for the contents they disseminate, even if they are not directly responsible for putting them online, because it is indeed the centralized and lucrative platforms such as Facebook or YouTube that are targeted. Thus, even if it is true that the possibility of using filtering mechanisms (which already exist for these platforms) can constitute an obstacle to freedom of expression, we can also question the very existence of such freedom on these platforms, which already subject their users to the growing influence of an algorithmic logic.

## 1.5. Recent extensions of the BCIS approach

Building on the pioneering work of Boyle, Lessig and Benkler on information and creative commons, several research programs have been conducted at the European level by researchers with close ties to the BCIS.

---

46 https://scinfolex.com/2018/09/15/la-directive-copyright-not-a-don't-go-for-internet-free-and-open/.

Although an explicit filiation has never been named as such, it is nevertheless very present in the questions and issues raised.

The first major research program that followed on from this work is called Communia. It is coordinated by Nexa, the Research Center for Internet and Society of the Polytechnic University of Turin (equivalent to the BCIS), and brings together 50 organizations (universities, libraries, archives, etc.) from the European Union, as well as from countries outside Europe such as the United States and Brazil. The BCIS is part of this network through the contribution of one of its founders, Richard Nesson. By proposing to shed light on the issue of the public domain in the digital ecosystem, this research program contributes to clarifying conceptual proximities and divergences with the emerging notion of information commons.

The second European program that we will discuss later is also in line with the work initiated by the BCIS jurists, because it raises the question of the need to deploy common elements at the level of network infrastructure to promote the deployment of a real cultural commons economy in terms of content. This is the netCommons Project[44], coordinated by the University of Trento (Italy), which is part of the European H2020 horizon. Five other European universities are involved, including CNRS researchers such as Mélanie Dulong de Rosnay. This program aims to study the conditions for deploying community digital infrastructures as new forms of common areas that can be used for the development of new technologies to become, in the long term, complementary networks, or even substitutes for the dominant infrastructure model based on a strong socio-economic domination in terms of services by Web giants.

### 1.5.1. *The digital public domain: the perimeter of cultural commons*

Communia is a thematic network on the Digital Public Domain (DPD) funded by Europe (2007–2009)[45]. Echoing James Boyle's pioneering

---

44 https://netcommons.eu/.

45 "COMMUNIA Thematic Network has been working for over three years at becoming a European point of reference for theoretical analysis and strategic policy discussion of existing and emerging issues concerning the public domain in the digital environment – as well as related topics, including, but not limited to, alternative forms of licensing for creative material; open access to scientific publications and research results; management of works whose authors are unknown (i.e. orphan works)." See: http://communia-project.eu/about.html.

approach to defending the public domain, this program advocates the proposal for a positive definition (from a legal point of view) of the public domain so that it can be protected from any attempt to "enclose" it. It reiterates the importance in the digital ecosystem of fostering open access to information resources. One of the objectives of this program is to make policy recommendations to strengthen the public domain in Europe in line with the European digital agenda.

Their main argument in favor of a positive recognition of the DPD is socio-economic. As Giancarlo Frosio points out, drawing on the work of economist Rufus Pollock[46], who founded the Open Knowledge Foundation[47]: "Value can be extracted from the structural and functional aspects of the public domain" (Frosio 2012, p. 10). Frosio takes up the theory of the economist Joseph Stiglitz, according to which knowledge is a public good producing positive externalities, which can disappear if markets seek to control knowledge excessively. A market that excessively privatizes information will be less efficient in the allocation of resources in society because the information facilitating this allocation will be harder to find. Most importantly, these digital public domain resources have a positive economic value. The use, or reuse, of public domain resources increases their economic value by offering a set of free resources that can give rise to innovation dynamics and new business models. For all these reasons, the protection of the DPD will also have a social value by allowing greater access to culture. We find here the liberal vision underlying Benkler's analysis in particular.

In this perspective, Communia's main conceptual contribution is to propose a definition of the wider public domain and to link it directly to the notion of cultural commons within their *Public Domain Manifesto*. The Digital Public Domain (DPD) is defined, in a broader sense, by two components:

---

46 Pollock, R. (2006). The Value of the Public Domain. UK Institute for Public Research Policy.

47 The Open Knowledge Foundation is a non-profit association under British law promoting free culture created in 2004. It provides technical tools such as CKAN, which enables the hosting of metadata associated with data catalogs, enabling governments, for example, to provide a catalog of their public data quickly and cheaply. It also offers legal assistance in the choice of licenses for open source content.

– the structural DPD, which includes all works that are outside the scope of copyright (facts, ideas, etc.) and those for which the term of protection has expired;

– the functional DPD, which groups together "voluntary commons", in other words resources that have been voluntarily made freely available by their authors and resources resulting from the reuse of copyrighted resources but which fall within the scope of exceptions (such as fair use).

These two components constitute our cultural heritage and thus form a kind of "global cultural commons" according to Mélanie Dulong de Rosnay and Juan Carlos de Martin: "The emergence and growth of an environnemental movement for the public domain and, in particular, the digital public domain, is morphing the public domain into the commons. The public domain is our cultural commons" (Dulong de Rosnay and de Martin 2012, p. 8). On the other hand, user rights are not the same in each case, and therefore governance is also different. The structural PD is a commons where all uses in terms of reuse are possible, whereas the functional PD is a commons built with rights of use that can be more or less permissive.

In terms of recommendations, the preamble to the presentation of their research program states that the European Digital Agenda is fully in line with their conception of the DPD by supporting the following principles: the digitization of the European cultural heritage with the support of the Europeana digital library, the necessary simplification and clarification of copyright on a European scale, in particular on orphan works, the promotion of cultural diversity and creative content in the digital[48] environment. Communia's main recommendations for strengthening the digital public domain, as set out in their *Manifesto*, are in line with the theses defended by BCIS jurists, among which: reduce the duration of copyright protection to

---

48 "In drafting these policy recommendations, COMMUNIA shares very much the vision of Neelie Kroes, European Commission Vice-President for the Digital Agenda, that "[c]ulture is the peak of human creativity and a source of collective strength" and "we want 'une Europe des cultures.'" The promotion of the public domain is empowering that "collective strength" and the European public domain is quintessential of "une Europe des cultures". The riches of digitization may multiply endlessly our cultural collective strength. However, new enlightened policy approaches and solutions are needed to reap the benefits of the present groundbreaking technological advancement. Again, the words of the European Commissioner Kroes powerfully convey the agenda of a modern digital Enlightenment that COMMUNIA aspires to propel with the help of the Commission". Citation from the Communia website, available at: http://communia-project.eu/final-report/annex-iii.html.

promote access to shared culture and knowledge, take into account the effects on the public domain of any change in the scope of copyright, any content that falls into the public domain in its country of origin must be recognized as belonging to the structural public domain in all other countries of the world, any false infringement of the public domain must be punished by law. Finally, it should be noted that several other European[49] projects, initiated during the same period, are in line with the Communia network's *Public Domain Manifesto*.

This creative revolution that the members of the Communia network are calling for also implies that socio-technological dimensions (and not only those related to the legal dimension of copyright) be taken into account. In the foreword to *The Digital Public Domain*, Charles R. Nesson, one of the founders of the BCIS and professor of law at Harvard, states that the most important recommendation made by the Communia network is the need to develop a digital registry of cultural content that allows any potential user to determine at zero cost which content is copyrighted and which is in the structural or functional public domain: "Seen from the perspective of users of the public domain, the greatest legal constraint on dissemination of public knowledge is from the threat of copyright litigation" (Nesson 2012, p. 12). Such a registry would allow everyone to be able to reuse content in a creative perspective without fear of legal action by potential rights owners. In this perspective, Communia proposes that each country, with the support of the Europeana digital library and the major European universities, initiate such a registry with, in a second step, the possibility of aggregating them in a global consortium.

We can note that the Creative Commons association has worked in this direction by working on the implementation of a search engine[50] that allows contents to be found easily under CC license or belonging to the public domain. It is defined as a tool for creators who not only want to discover but also reuse free resources with ease and confidence:

---

49 The Europeana Foundation, which published the Charter of the Public Domain in 2008; the LAPSI project, which brings together a network of reflections on the access and reuse of public sector information in the digital environment; the Rightcom project on the economic and social impact of the public domain; the DʹARIAH project (Digital Research Infrastructure for the Arts and Humanities) with the aim of fostering digital research between the humanities and the arts; the ARROW project aimed at finding ways to clarify and easily identify the state of copyright on works; and finally the DRIVER project, which proposes to build an infrastructure and a search engine for all open scientific communications.

50 https://ccsearch.creativecommons.org/.

The vision centers on reuse – CC will prioritize and build for users who seek to not only discover free resources in the commons, but who seek to reuse these resources with greater ease and confidence, and for whom in particular the rights status of these works may be important. This approach means that CC will shift from its "quantity first" approach (front door to 1.4 billion works) to prioritizing content that is more relevant and engaging to creators.[51]

| Source | Domain | # CC Licensed Works |
|---|---|---|
| PhyloPic | http://phylopic.org | 3,463 |
| Flora-On | https://flora-on-on.pt | 5,501 |
| Rawpixel | https://www.rawpixel.com | 6,355 |
| Thorvaldsens Museum | http://www.thorvaldsensmuseum.dk | 8,912 |
| Culturally Authentic Pictorial Lexicon | http://capl.washjeff.edu | 15,142 |
| Animal Diversity Web | https://animaldiversity.org | 15,554 |
| McCord Museum | http://www.musee-mccord.qc.ca/en | 21,872 |
| World Register of Marine Species | http://www.marinespecies.org | 23,716 |
| Thingiverse | https://www.thingiverse.com | 29,624 |
| Rijksmuseum | https://www.rijksmuseum.nl/en | 29,999 |
| Cleveland Museum of Art | http://www.clevelandart.org | 32,643 |
| Sketchfab | https://sketchfab.com | 37,903 |
| Brooklyn Museum | https://www.brooklynmuseum.org | 61,503 |
| Museums Victoria | https://museumsvictoria.com.au | 85,575 |
| Digitalt Museum | https://digitaltmuseum.no | 266,672 |
| DeviantArt | https://www.deviantart.com | 271,362 |
| SVG Silh | https://svgsilh.com | 276,966 |
| Metropolitan Museum of Art | https://www.metmuseum.org | 500,738 |
| Geograph Britain and Ireland | https://www.geograph.org.uk | 1,244,387 |
| Behance | https://www.behance.net | 6,479,672 |
| Wikimedia Commons | https://commons.wikimedia.org | 23,749,024 |
| Flickr | https://www.flickr.com | 349,021,635 |

**Table 1.1.** *List of content provider sites under open license*[52]

51 https://creativecommons.org/2019/03/19/cc-search/.

52 https://ccsearch.creativecommons.org/.

To date, it provides access to content in the form of open licensed or public domain images from open APIs and the common crawl database. Table 1.1 provides a list of data providers.

Within these 22 data vendors are very disparate entities. We find museums, such as the Rijksmuseum (Amsterdam) or the Metropolitan Museum of Art (New York), which have a proactive policy in terms of reusing their digital heritage, amateur platforms for "royalty-free" photos with the largest provider of content in Creative Commons format, Flickr (which alone accounts for more than 90% of the content accessible by this search engine), platforms such as Behance dedicated to designers, amateurs and professionals, without forgetting Wikimedia Commons (which accounts for 6% of content).

At this stage, it might be interesting to have more information on how the algorithm underlying this search engine was built and the actual uses it has developed. Let us recall that this last point is an essential aspect on which the arguments of the defenders of a cultural commons economy are based. But how can such a study be implemented?

Indeed, in the absence of systematic identification of the users of this search engine, it seems difficult to us to realize it. Finally, there is also the question of the visibility of this search engine in the space of the Web, where the search for information is dominated by commercial search engines with Google in a quasi-monopolistic situation.

## 1.5.2. *Network infrastructure as a commons*

Yochai Benkler was one of the first to stress the importance of network infrastructure as an indispensable lever for the deployment of cultural commons at content level[53].

In a way, this netCommons research program aims to explore this issue more closely by starting with an in-depth field analysis of case studies of emerging community networks:

---

53 See Benkler, Y. (1998). Overcoming agoraphobia: building the commons of the digitally networked environment. *Harvard Journal of Law and Technology*, 11(2), Winter.

Community networks (CNs) started appearing in the 1990s, as the internet was growing in popularity. They have been called by many names: free networks, alternative telecom providers, do-it-yourself internet service providers (ISPs), etc., but basically, it is about managing telecommunications as a commons, that is, a resource produced and maintained collectively, rather than held privately.[54]

These networks are defined as local telecommunication infrastructures set up according to a bottom-up logic by groups of people (the community) allowing them to connect to the Internet and offering digital communication services.

The particularity of these community networks is also to provide a guarantee of transparency on the flow of personal data and their reuse. Several community networks are thus examined. A certain number of them have been built to cope with the lack of a possible connection to the Internet network in certain territories. For example, Guifi.net was created in 2004 in the Spanish region of Osona to solve the difficulties of broadband access in these rural areas. In terms of technical infrastructure, this network has established connections through Wi-Fi routers and a set of geographically close people who decided to deploy their own network by interconnecting different nodes (houses, offices, libraries). Today, it is considered to be the largest citizen Wi-Fi network in the world. As of December 2016, it included more than 32,500 active hubs, most of them in Catalonia, as well as many others in Valencia, the Balearic Islands, Madrid, Andalusia, Asturias, and the Basque Country. Many of these networks are projects initiated by hackers belonging to associative networks carrying the values of free culture and DIY (Do It Yourself).

In this research program, the notion of the common is explicitly invoked to refer to the fact that these networks are built and governed collectively (rather than privatized through the commercial entities that are Internet service providers). The challenge of this research program is to study the characteristics of these so-called "collective" modes of production and

---

54 Dulong de Rosnay, M., Treguer, F. (2018). Telecommunications Reclaimed: a hands-on guide to networking communities. *Internet Society.* Available in CC BY on www. net.commons.eu.

governance in order to identify the invariant forms, as well as those that over time prove to be the most efficient. The researchers of the netCommons project began this reflection by taking as a starting point the conceptual model of common pool resources of the contemporary theorist of the commons, Elinor Ostrom:

> The theoretical framework of the commons in general, and of commons-based peer production in particular, is a reference for the development, management, and scientific analysis of CNs (community networks) ... the underlying principle behind CNs is the firm conviction that the CPR framework presents the optimal way to run a network, as a critical resource for the development and sustainability of a community. CPRs were studied in depth by E. Ostrom.[55]

These community networks are an illustration of what they call artificial digital commons and are defined as rival and non-exclusive resources. The governance of these commons is based on the action of several groups of actors[56] each characterized by specific bundles of rights. These artificial digital commons fulfill an essential function, which is to provide citizens with the means to build and participate in a self-organized way in social connection and access to knowledge and means of communication. The commons approach allows this research program to assess in a comparative perspective which community networks are the most sustainable.

The exploration of this infrastructural dimension seems essential to us. It is unquestionably one of the essential axes of a research program on the

---

55 The first deliverable of this project is very instructive in this respect, as it puts into perspective, while adapting it, Ostrom's theory in the case of digital commons: www. netcommons.eu/sites/default/files/d1.1_reportexistingcn_dlv.pdf.

56 "The volunteers, the initiators of the project, due to their lack of economic interests, are responsible for the operation of the tools and mechanisms of governance and oversight. The professionals bring in quality of service, and their customers bring the resources which make the ecosystem economically sustainable. Public administrations are responsible for regulating the interactions between the network deployment and operation, and public goods, such as public domain occupation. All participants that extract connectivity must contribute infrastructure, directly or indirectly, and can participate in the knowledge creation process": www.netcommons. eu/sites/default/files/d1.1_reportexistingcn_dlv.pdf, p. 22.

cultural commons[57], because it raises the question of the articulation between the nature of networked infrastructure and the modes of production and circulation of knowledge and culture. A community network infrastructure as studied here is not a necessary condition for the deployment of a cultural commons economy, which, as has been shown, has been able to emerge and deploy itself within a commercial network infrastructure. However, if the conditions for cohabitation with the commercial economy actors prove difficult over time, preventing their growth or even their very existence, then we may wonder whether such infrastructures do not constitute fortresses likely to guarantee their existence in the medium term.

### 1.5.3. *Remuneration of volunteer contributors*

A final avenue for extending the pioneering work of the BCIS addresses the issue of compensation for voluntary contributors to the production, individually or collaboratively, of cultural commons. The cultural commons economy is based on the cohabitation of non-market and hybrid platforms, each of which hosts diverse content produced by volunteer contributors, individually or collectively. Following the approach of the BCIS pioneers, two researchers associated with the BCIS, Primavera De Filippi[58] and Samer Hassan[59] (2014), have begun to reflect on the advantage of remunerating voluntary contributors in the case of platforms based on collective contributions such as Wikipedia, Creative Commons, CouchSurfing or Open Street Map, which they call Commons-Based Peer Platforms (CBPP) in reference to Benkler's pioneering approach. We see here that we are going beyond the strict framework of intellectual works, and therefore beyond our narrow understanding of the notion of cultural goods. Nevertheless, their analysis seems interesting to present here, as it places the question of perenniality and the enrichment of the commons in a broader perspective encompassing all existing forms.

---

57 We refer the reader to the first deliverable of the project, available online at www.netcommons.eu/sites/default/files/d1.1_reportexistingcn_dlv.pdf.

58 CNRS Researcher at CERSA (*Centre d'étude et de recherche en sciences administratives et politiques*) of the Université Paris 2. She is also associated with the BCIS. Her work focuses on the legal implications of distributed architectures such as blockchains and how they could contribute to new forms of governance.

59 Professor at the Complutense University of Madrid. He is also associated with the BCIS.

De Filippi and Hassan's approach focuses on the analysis of Commons-Based Peer Platforms (CBPP) as defined by Benkler. They defend the idea that it might be relevant to construct a metric evaluation of each contributor's social value to the commons platforms, as this could encourage individual incentives to contribute and thus promote the enrichment of the commons. This estimate could constitute a kind of reputational capital that the contributor could also bring to bear on the market.

This individual indicator of social value is not part of a market price system because it could be counterproductive. Indeed, introducing monetary remuneration linked to the influence of each person's contributions to these CBPPs could introduce undesirable individual opportunistic behaviors that could jeopardize this sharing ecosystem. If some contributors are paid directly according to their production, then others will no longer want to contribute for free; moreover, since money is scarce, this may lead to strong individual competition instead of collaboration. Finally, another undesirable effect would be to encourage contributors to focus on projects that are considered the most remunerative. This individual indicator of social value is also not based on a non-material reward system (reputation, administrative rights, privileges) because this would not improve the voluntary contributor's (economic) situation. It would be at the interface between the non-market sharing ecosystem and the market by providing rewards that are non-transferable (not based on a monetary equivalent), but that the market would nevertheless be able to recognize (and therefore internalize). From this perspective, De Filippi and Hassan propose an alternative metric of value to that provided by the market (the price system) that is based on two principles:

– social value emerges from within the network of actors contributing to CBPPs;

– social value is subjective, that is, based on the cross-perception of contributors to other CBPPs.

This indicator is constructed in several stages. The first consists of constructing a quantitative social value indicator associated with each CBPP community. The second step is to assess the individual social value of each estimated contributor in a common metric, the Sabir.

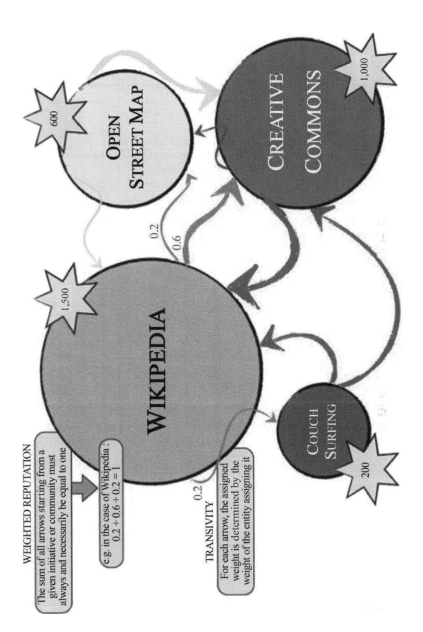

**Figure 1.1.** *Example of social values for different Commons-Based Peer Platforms. For a color version of this figure, see www.iste.co.uk/pelissier/commons.zip*

Let us take a look at some elements of their demonstration. At the beginning is an ecosystem composed of a predefined number of CBPPs (e.g. Wikipedia, Creative Commons, Europeana, Flickr). Each community allocates vouchers to the other CBPPs of their choice by assigning them a weight (knowing that the sum of these relative weights must be equal to one). By applying a popularity principle similar to pagerank, a social value can be calculated for each CBPP (e.g. Wikipedia: 5,000, Creative Commons: 2,000).

Second, each contributor to the CBPPs receives non-transferable tokens from the governance of the various CBPPs. Each community is free to decide the number of tokens it produces and how it distributes them, knowing that in each case this is based on an internal reward mechanism based on gratitude and appreciation. In order to make the value of everyone's contribution comparable, the value of each token is translated into a common denominator of value, the sabir: "Sabir can essentially be regarded as a proxy for value in the Commons-based economy. Just like prices do in the market economy, Sabir allows for individual contributions to be assessed and compared according to a common denominator value (which remain distinct from the market)" (De Filippi and Hassan 2016, p. 11). For example, Wikipedia could award its tokens according to the various possible contributions: creation of articles, revision, etc. (De Filippi and Hassan 2016, p. 11).

The value of a token for a community associated with a commons is the ratio between the social value associated with that community (CBE) and the total number of tokens produced by that community. It can be noted that it is therefore not in the interest of the community to produce too many tokens, as this will lower the value of the token it produces.

Like the price system used as a proxy for economic value, the sabir system as a proxy for the social value of the commons is based on a decentralized logic with no central authority responsible for assigning the value of resources, which will spontaneously contribute to the emergence of positive externalities. In fact, this system for estimating the social value of commons platforms and each contributor's social value is of interest above all because it constitutes a kind of common language, a form of interface between the universe of commons and that of the market that allows everyone to be in a mutually beneficial relationship: voluntary contributors can put their estimated social value to use on the market and receive (non-monetary) rewards in the form of a free offer of goods or services from economic actors who wish to value this type of action. It is assumed that the more companies recognize this new system of social evaluation, the more it will allow everyone to increase the time devoted to contributing to the commons:

> Over time, a positive feedback loop will therefore be established, as market entities that support (or sponsor) the commons will gain reputation within the commons ecosystem. This might, ultimately, bring more and more market players (whether or not they are themselves CBPP contributors) to purchase their goods or services on the market, knowing that, by doing so, they are also helping the commons (De Filippi and Hassan 2014, p. 12).

A number of questions remain unanswered when reading this thesis[60]. However, its merit is that it opens up a central question, the remuneration of contributors, and proposes a formalization based on the construction of a social value indicator. We shall see in the next chapter how this question is also a central preoccupation of the representatives of the Ostromian approach to the commons. But it will become apparent that the perspective adopted is fundamentally different from the one presented here.

---

60 Certain points seem to us to need to be clarified, in particular on the modalities of attribution of vouchers by each community of commons to the other commons. Indeed, it can be assumed that this is the responsibility of the governance of these commons and that specific rules (such as voting) should prevail. But nothing is said in this article to that effect. Similarly, in the construction of the social indicator, the social value of a common at time $t$ being based on the social value of other commons, defined at time $t - 1$, we may wonder how the social value of the common or commons created at time $t = 0$ is defined.

# 2

# The Ostromian Approach to the Knowledge Commons

Elinor Ostrom was the first woman to receive the Nobel Prize in Economics, in 2009, for her analysis of the economic governance of the original institutional forms of common land resources. We pointed out in the introduction that this work, which began in the 1960s at UCLA and continued until her death in 2012, focused almost exclusively on the governance of certain types of natural resources (fisheries, forests, pastures, etc.). When James Boyle invited her to the Duke symposium in 2003, she was given the opportunity to extend her reflection to the field of intangible resources (information, culture, knowledge). Two major publications followed, both written in collaboration with Charlotte Hess[1] (Hess and Ostrom 2003, 2007).

These various writings do not constitute a finalized and completed theory on the knowledge commons. Rather, they consecrate the creation of a research program on this theme that they consider fundamental. In this perspective, they lay down some essential milestones, at a theoretical and

---

1 The latter is not an academic but the director of the research library of the *Workshop in Political Theory and Policy Analysis* at Indiana University, led by Elinor Ostrom and her husband and fellow academic Vincent Ostrom. This also most probably explains the interest in their reflections focused primarily on what we now call the scientific commons. After their joint contribution in 2007, Charlotte Hess continued their reflection in several articles, of lesser resonance, but which are also important because they are in line with the influence of Ostrom's work: Hess, C. (2008). Mapping New Commons. *The Twelfth Biennial Conference of the International Association for the Study of the Commons*, Cheltenham, Great Britain, July 14–18; Hess, C. (2012). The Unfolding of the Knowledge Commons. *St. Anthony's International Review*, 8(1), 13–24.

methodological level, to apprehend, from an empirical point of view, the study of the knowledge commons in specific contexts. This notion has much broader outlines than that of the cultural commons, which is only one of their manifestations. In the book on the knowledge commons that they have edited, Hess and Ostrom (2007) devote a significant part of their reflections to the question of open archives as a possible illustration of knowledge commons in the digital ecosystem that could act as a barrier to the different forms of enclosures that are manifested in the digital editorial ecosystem. Representatives and advocates of open access (OA), such as the philosopher Peter Suber and Nancy Kranich, former president of the Association of American Library Associations, participated in this book. By strategically linking the OA movement with the commons movement, this book opens up a new avenue for the possibility of rethinking the current reconfigurations of the scientific editorial ecosystem under the prism of the institutional approach of the commons. This is the door opened by Hess and Ostrom that we propose to open wide by tracing some directions to follow not only to reveal the specific nature of open archive[2] platforms as potential scientific commons, but also by studying the conditions of their cohabitation with the platforms proposed by traditional scientific publishers.

This approach reveals in the background of these divergent editorial logics the question of the platforming of the dynamics of production, distribution and circulation of knowledge and the foundations of the knowledge economy more generally. It is a question that runs through a whole other contemporary research program that is interested in the different manifestations of this return of the commons in the digital ecosystem, as a continuation of Ostrom's work. We mention here the work initiated in France by the economist Benjamin Coriat within the framework of several successive research programs on the knowledge commons. Several intellectual figures revolving around the question of the economics of collaborative platforms are associated with this work. While some of them, such as Michel Bauwens, are known for belonging to the militant sphere of the commons, others belong to the community of researchers working in the field of social and solidarity economy, at university level as well as in the

---

2 We use the term platform in the following sense as defined in the French Digital Republic Act: "Activities consisting of classifying or referencing content, goods or services offered or placed online by third parties, or of bringing together, by electronic means, several parties with a view to the sale of a good, the provision of a service, including on a non-remunerated basis, or the exchange or sharing of a good or service" (Article 22).

field itself. Beyond an initial work of identification of collaborative platforms that could be considered as commons, it is also a question of studying the conditions of their deployment and their sustainability in a universe dominated by platforms that rely on opposite modes of value exploitation and forms of governance. We have named them social commons because of their claimed proximity to the foundations of the social and solidarity economy. Even if the platforms studied in these various works are primarily service economy platforms, some illustrations are also examined in the field of culture.

## 2.1. Ostrom's original theory of the land commons

### 2.1.1. *An institutional definition of the commons*

In her lecture at the Nobel Prize in Economics[3], Elinor Ostrom recalled that her research on the problems of collective action faced by individuals when using common resources began with her doctoral work in the early 1960s, when she worked on water management in Southern California. She later continued her reflection with her husband and fellow academic colleague Vincent Ostrom at Bloomington University in Indiana. In the 1970s, they jointly created an interdisciplinary workshop entitled *Political Theory and Policy Analysis*, which proposed to evaluate, based on multiple empirical studies, the institutional arrangements that regulated natural resources in different ecological and socio-economic contexts.

Their work took on an unprecedented scale in the 1980s when they were associated with the program launched by the National Research Council (NRC). This enabled them to bring together all the work done worldwide on this issue by different groups of researchers, anthropologists, historians, sociologists, politicians, etc. For Benjamin Coriat (2013), the Annapolis conference (1985), during which all the empirical and theoretical advances initiated by the NRC program were presented, constitutes the moment when the program on commons took on a new impetus and what he called "the great return of the commons". During this period, at the international level, the World Bank initiated so-called "structural adjustment" programs in line with the doctrine advocated by the Washington Consensus, which legitimized liberal policies at the economic level. Thus, all support policies

---

3 *Beyond Markets and States: Polycentric Governance of Complex Economic Systems* conference, available at: www.nobelprize.org/prizes/economic-sciences/2009/ostrom/lecture/.

in developing countries had to be based on an incentive to privatize resources so that a supposedly efficient logic of market mechanisms could be put in place. However, there was strong concern among some key development actors about the failures of aid policies aimed at promoting agricultural productivity in various countries in the tropics.

On the research side, the results of Elinor Ostrom and her team revealed that many natural resources were managed neither by the State nor by the market but by self-managed communities of individuals. This proved, in the long term, to be perfectly efficient in ensuring the survival of the populations that live off them and for the preservation of the resource itself. Such an observation thus counterbalanced the liberal ideology that made the market the only efficient institution for the management of rare resources. However, Elinor Ostrom never defined this governance commonly as a model to be applied everywhere and for any common resource. First of all, situations of failure in such modes of governance have also been observed. Secondly, this form of institutional governance only applies to certain types of common resources and in certain contexts. The common is not thought of as an alternative to the market. Rather, the research program it created aims to provide a better understanding of the relationship between the resources studied (land, pasture, forests, water resources, etc.) and the associated ownership regimes. The institution of the commons cannot be interpreted as a general principle for reorganization of socio-economic order. It is necessary in certain situations, for certain specific goods, but does not call into question the efficiency of markets and the State as forms of governance. As Dardot and Laval point out, "Elinor Ostrom is not anti-capitalist, nor is she anti-state. She is liberal. Favorable to institutional diversity, she trusts in the freedom of individuals to invent for themselves, outside of any governmental constraint, the contractual agreements that benefit them" (Dardot and Laval 2014, p. 155, author's translation). The unanswered question that Ostrom did not answer herself is how these resources governed under a commons mode can endure and develop in the dominant organizational modes of the market and the State.

## 2.1.2. A questioning of the "tragedy of the commons"

For a very long time, the principle of shared ownership has been strongly criticized, as Guibet Lafaye (2014) points out. This "economic disqualification of the commons", which dates back to Aristotle, finds ardent defenders in the contemporary era. The theory of a "tragedy of the commons" put forward by the biologist Garett Hardin (1968) symbolizes the environmental degradation to be expected when several individuals share a limited resource. Each individual, driven by his or her personal interest, will be encouraged to overexploit the resource, which runs counter to its preservation over time. This theory was also defended by the economist Mancur Olson in his 1965 book *The Logic of Collective Action*, as Elinor Ostrom recalls: "Olson challenged the presumption that the possibility of a benefit for a group would be sufficient to generate collection action to achieve that benefit" (Ostrom 1990, p. 6). It was taken up a few years later by the neo-institutionalist economists Demsetz and Alchian (1973). For the latter, the communal property regime associated with common resources is based on a "first come, first served" principle that inevitably leads to situations of social dilemma preventing the emergence of a cooperative solution. In turn, arguments in terms of cost (costs of negotiation to agree on the exclusion of those who do not respect the rules, for example) and efficiency (problem of individual incentive to cooperate) are put forward. The only solution, according to them, to get out of it is to impose a private and exclusive property regime. In other words, it is a question of internalizing the negative external effects of a communal property regime. Another possible solution would be to entrust the management of these resources to the State, which would levy taxes and define access rights. In the background of these different approaches is the famous problem embodied in the model of the prisoner's or free rider's dilemma, which can be summed up as follows, as Elinor Ostrom insistently pointed out: "At the heart of each of these models is the free-rider problem: whenever a person cannot be excluded from the benefits that others provide, each person is motivated not to contribute to the joint effort, but to free ride on the efforts of others. If all participants choose to free ride, the collective benefit will not be produced" (Ostrom 1990, p. 6).

Elinor Ostrom did not question the possible existence of free riding behaviors that potentially threaten these types of resources due to their non-exclusive nature. Nor did she question the dominant assumption of neoclassical models that individuals are driven by their personal interest in the choices they face in such decision-making contexts. She even acknowledged that this problem clearly arises for the natural resources that are at the center of attention. She defines as "common pool resources" (CPRs) any system of resources, natural or man-made, that is large enough to make it costly (or even impossible) to exclude potential beneficiaries. It is precisely this property of non-exclusivity that threatens resources with depletion, all the more so since CPRs are made up of a "pool" of resource units (a fishery and the fish, a forest and its plants, for example) which are also called rivals because their consumption leads to a reduction in the quantities available to others. This dual property of non-exclusivity and rivalry is what characterizes common land resources.

On the other hand, contrary to Hardin, Olson, Demsetz and Alchian, for Elinor Ostrom, the recognition of this social dilemma does not necessarily imply that the only effective governance is the market (through the privatization of the resource) or the State (through the establishment of public property). Hardin's allegory is based on a very restrictive experimental model that cannot correspond to all real cases. Indeed, it assumes that individuals make independent decisions by focusing primarily on their short-term personal gains. However, Ostrom's contribution is precisely to show that from the moment that we introduce, in the modeling of decisions, the possibility for these same individuals to communicate through face-to-face discussions, then they may be able to develop and comply with cooperation standards. She deliberately cited the contributions of the evolutionary game theory, which models game situations corresponding to the prisoner's dilemma model in the context of repeated interaction with individuals who, being endowed with limited rationality, learn in the course of interactions to conform to rules and norms of cooperation, reciprocity in particular. These approaches show that it can be quite rational to cooperate in such contexts[4].

Ostrom's second criticism concerns the conception of ownership attached to these common resources. In particular, Hardin retains a conception of communal property in terms of a negative community, that is, resources that are freely accessible and whose use is open to all. Behind this conception lies the ancient qualification of the common thing, inherited from Roman law and set out in article 714 of the French Civil Code, as Judith Rochfeld points out: "There are things that belong to no one and whose use is common to all" (Rochfeld 2014, p. 357, author's translation). However, this view focused on goods that, by their nature, could not be appropriated and were considered (at the time) to be available in abundance, such as water or air, and not on natural resources such as forests or fisheries, which are scarce resources. For Ostrom, however, this conception of communal property did not apply in the

---

4 Ostrom cited the pioneering work of Axelrod (1981, 1984) and Kreps and Wilson (1982).

case of the CPRs she studied. Not only are these resources not accessible to all, but the rights of use are not necessarily the same for all authorized persons. She introduced a conception of communal property in terms of bundles of rights.

### 2.1.3. *Communal property as a bundle of rights*

Numerous empirical studies reveal that common pool resources that present themselves as CPRs are often regulated by a complex system of shared ownership where the different stakeholders of the common pool resource (still defined as commoners) share rights and obligations regarding access and removal of the resource. This is one of the main results of Ostrom's work and that of her team based on investigations carried out in different parts of the world: communal forests in Japan and Switzerland, irrigation systems in Spain and Sri Lanka, water tables in California, and fishing in Turkey or Scotland.

Ostrom's work is based on a conception of property defined not as the relationship of an individual with a thing, but as a bundle of relationships existing between individuals concerning a thing. More precisely, she defined property as a bundle of rights (and obligations) attributed to individuals, the function of which is to regulate inter-individual relations around the use and management of natural or man-made resources. This approach was not entirely new. For Fabienne Orsi (2015), it was first enunciated by the American institutionalist economist John Commons and the North American legal current known as legal realism dating back to the end of the 19th century. This legal realism is part of a progressive movement which, at the time, was opposed to the dominant liberal vision advocating "laissez-faire" in terms of economic regulation and defending a conception of property as an immutable natural right.

This conception of property as a bundle of rights was set out in an article co-written with the jurist Edella Schlager (Ostrom and Schlager 1992). Their aim was to show how common resources are regulated by complex and highly diversified property regimes that resemble a distribution of rights (and obligations) between partners (commoners) associated with the exploitation of the resource. Thus, to each possible combination of these rights will correspond different types of commons.

There is no correlation between a type of good (public, private, common) and a specific property regime. In particular, a common resource is not intrinsically common in the sense of Ostrom. As Olivier Weinstein points out, "the important thing for Ostrom was not to identify a few major forms of property, such as private property, common property or public property, but to show how a specific regime can be constructed for each particular situation. What matters is institutional diversity" (Weinstein 2015, p. 77, author's translation).

Five types of rights define the ownership of a CPR (Ostrom and Schlager 1992); they are divided into two categories:

– the so-called operating rights defining the actions that are authorized in relation to the use and the modalities of appropriation of the common resource:

- "access: the right to enter a defined physical property",

- "withdrawal: the right to obtain the 'products' of a resource (e.g. catch fish, appropriate water, etc.)";

– individuals with these two types of rights may also have other rights to the common resource (but this is not mandatory). These other types of rights fall under the regulation of collective actions and, in particular, the definition of collective rights. There are three of them:

- "management: the right to regulate internal use patterns and transform the resource by making improvements",

- "exclusion: the right to determine who will have an access right, and how that right may be transferred",

- "alienation: the right to sell or lease either or both of the above collective choice rights".

Individuals who own management rights have the authority to determine how, when and where appropriation can take place. Owning exclusionary rights gives the authority to define who has the right to access the common resource. Finally, the ownership of a right of alienation means that we have the authority to sell or transfer all or part of the collective rights we hold (management and exclusion).

Starting from this typology in terms of bundles of rights, Schlager and Ostrom wanted to show that there are several types of "owners" of a common resource according to the rights they hold. They take the example of salmon fishing grounds in different parts of the world where commoners have property rights ranging from a simple right of use to all five rights. Communal property regimes are those where commoners have at least collective rights in terms of management and exclusion.

Finally, it is important to make a small input here by proposing a comparison between this approach to ownership and the ownership regime defined by open licenses such as the Creative Commons. Indeed, we could be tempted to see CC licenses as the embodiment in the field of knowledge of Ostrom's shared ownership design, but that would be a gross error. This does not mean that their comparison is irrelevant. Ostrom's approach remains a very relevant reference model to put into perspective precisely what constitutes the singularity of the decision-making context in the context of shared cultural resources in the digital ecosystem.

Remember that CC licenses are so-called open licenses that introduce a concept of shared ownership based on the possibility for users to copy creative content and share it. They therefore have a right to use and appropriate these resources. However, the first significant difference concerns the resource's degree of openness, in other words its accessibility. In fact, the users of the natural resources studied by Ostrom belong to a community whose outlines are well defined and, within it, the commoners do not necessarily have the same rights of use. Outside this community, it is not possible to have access rights. This is what distinguishes a common resource of the CPR type from a public good whose access is open to all without distinction. However, in the case of Creative Commons licenses, users do not belong to a community with delimited outlines: every person has a right of use in regard to the resources protected by such licenses and the rights of use apply equally to every user (depending on the rights conferred by the CC license that is chosen). If the principle of openness replaces the principle of exclusion, it is also because the type of common resource eligible for the application of a Creative Commons type license does not have the same properties as the common resources studied by Ostrom. In particular, knowledge is considered a non-rival resource, unlike land resources. The dilemma is therefore not primarily the free rider problem, since the consumption of a non-rival resource does not reduce the amount available for others. This leads us to another question, which we will not answer right away, which is whether open licenses of the CC type constitute the only legal regime applicable to the governance of a commons in the cultural field.

## 2.1.4. *An institutional approach to the self-organization of common resources*

Ostrom and her team's objective was to understand how a group of individuals in a situation of interdependence with respect to the use of CPR could organize themselves and derive mutual benefits when their immediate interest would dictate them to adopt opportunistic free-riding behaviors. Thus, Ostrom conducted numerous empirical studies seeking to understand how the community associated with such CPRs governed and coordinated itself, as well as how the rules that ensured their reproduction were born and formed. It is therefore an economic approach known as institution-based, because it focuses on the nature of the social rules and norms that condition individual decisions in a context of social interaction and questions their emergence.

The decisive contribution of these studies was to reveal that the rules governing the organization of CPRs are implemented by the commoners themselves, in a dynamic of interaction and mutual learning, in a context of uncertainty, and without the intervention of an external authority to ensure that they are respected. The conditioning force of these rules results precisely from their emergent character. In the course of interaction dynamics, commoners learned to respect certain rules that encourage them to cooperate because they realized that it was in their interest to do so in a complex, uncertain environment, where they did not have all the information they needed to make a decision that they considered optimal. This is not a matter of benevolence. Cooperative behavior is based on the correspondence between individual and collective interest. Commoners are rational individuals, but being in a situation of uncertainty, their rationality dictates that they implement contingent strategies and not strategies that are independent of the choices of others. Such a proposition had already been put forward in the field of evolutionary game theory: "The contingent strategy that has been the object of the most scholarly attention is tit for tat in a two-person game in which an individual adopts a cooperative action in the first round and then mimics the action of the opponent in future rounds (Axelrod 1981, 1984)" (Ostrom 1990, p. 36). Understanding the situational variables that can influence the decision-making space is an indispensable step in understanding the learning dynamics that give rise to the emergence of behavioral norms such as the principle of reciprocity.

Among these rules, Ostrom showed the decisive importance of the possibility given to commoners to communicate. Indeed, it is only on this condition that they can learn to mutually respect certain social norms (such as mutual trust), which in turn promotes, through mutual learning, the emergence of institutional rules relating to the uses of the resource units of the CPRs that will ensure their preservation. The possibility of communication between commoners is fundamental to the success of collective action. The possibility of face-to-face discussion promotes trust and the emergence of common rules of cooperation. It is therefore at the very foundation of successful collective management. In such a context, the commoners will develop, in a relatively autonomous manner and through a process of trial and error, systems of rules or more precisely institutional[5] arrangements, defining in particular the rights of access and appropriation of the common resource.

While commoners have learned over time to understand that it is their interest to commit to the rules, this does not preclude the possibility that uncooperative and opportunistic behavior may occur in the future. Yet, Ostrom's empirical work has revealed that some governance structures are proving to be more robust than others. The institutional arrangements built by user communities are never the same from one common resource to another, as they depend on the purposes attributed to the common. While there is no single model that can be applied in any common pool resource context, Ostrom defined what she called "design principles", which refer to a set of characteristics shared by all strong institutions to ensure the sustainability of common resources. The first three principles identified are as follows:

– the access and ownership rights of commoners must be clearly defined as well as the physical limits of CPRs;

– the benefits must be proportional to the costs incurred;

---

5 Here is the definition of institutions for Ostrom: "'Institutions' can be defined as the sets of working rules that are used to determine who is eligible to make decisions in some arena, what actions are allowed or constrained, what aggregation rules will be used, what procedures will be followed, what information must or must not be provided, and what payoffs will be assigned to individuals dependent on their actions" (Ostrom 1990, p. 51).

– most individuals affected by business rules can participate in their modification (they therefore have collective rights).

CPRs that are characterized by these three design principles generally produce good rules. But the presence of good rules does not necessarily imply that commoners will always follow them. Robust CPRs are in fact based on a set of other interrelated principles that consist of the establishment of monitoring rules and sanctions. Governance rules therefore address not only access and ownership, but also oversight mechanisms for conflict resolution, which are essential to ensure that any deviation from the collective interest is sanctioned. Here are the four design principles that ensure the robustness of a commons:

– procedures must be put in place to make collective choices;

– rules of supervision and monitoring must exist;

– graduated and differentiated sanctions must be applied;

– conflict resolution mechanisms must be instituted.

What should be retained from this synthetic overview of Ostrom's approach is that not all common resources are intrinsically common. While they are all *a priori* eligible to become so because of their economic properties (low exclusivity and strong rivalry), they will only become so if other conditions are met, which relate to the nature of the rights distributed among the stakeholders and the nature of the governance.

## 2.2. The knowledge commons: Hess and Ostrom's approach

### 2.2.1. *The singularity of information common pool resources (CPR)*

After this digression, which is essential to identify the foundations and outlines of Elinor Ostrom's commons theory, we now propose to focus on extending her approach to common information resources. In 2001, Elinor Ostrom and Charlotte Hess were invited to participate in the Public Domain Conference organized by James Boyle at Duke University School of Law. The conference brought together lawyers from the Berkman Center such as Lessig and Benkler, artists, environmental system specialists, archivists and computer engineers. In the foreword to the special issue of *Law and Contemporary Problems* presenting papers from this conference, James Boyle explains that he had invited Elinor Ostrom and Charlotte Hess to assess the extent to which their approach could be applied in a meaningful way to what he calls the intellectual commons: "One of our goals in organizing the conference was to turn Ostrom and her distinguished collaborator Charlotte Hess on the intellectual commons with the goal of discussing the applicability of their ideas to this new realm, and perhaps of producing a similar matrix of types of commons and strategies of management" (Boyle 2003b, pp. 1–2). According to him, they achieved a real "tour de force" by developing, in clear and understandable language, a methodology that can be applied to knowledge commons in the field of scientific communication in particular.

The reflection originally focused on natural physical goods was thus later extended to knowledge in an article in 2003, following their participation in the Boyle conference (Hess and Ostrom 2003), and later in a collective work entitled *Understanding Knowledge as a Commons, from Theory to Practice* (Hess and Ostrom 2007). This book is the culmination of a series of intellectual encounters on the conditions of adaptation of their environmental approach to the information domain. In the foreword of their book, they talk about their participation in the conference on the public domain organized by James Boyle at Duke University in 2001, and the one they organized in 2004, "Scientific Information, Digital Media and Commons," at Indiana University. While highlighting the remarkable work of the jurists who raised the alarm about the many threats of enclosure on knowledge relevant to the public domain, Hess and Ostrom were more nuanced in their interpretation of the notion of the commons in the information field and its often too rapid assimilation into the public domain[6]. The public domain refers to a body of knowledge taking very different forms that is not or no longer subject to copyright. This means that the uses that can be made of it are not subject to any regulation. Yet, in a land commons, commoners have rights of use in terms of access and appropriation that are delimited by rules (formal or conventional). They cannot do what they want. The space of freedom is more constrained.

Hess and Ostrom retained a broad definition of knowledge that led them to consider the expression information commons and knowledge commons interchangeably: "Knowledge refers to all intelligible ideas, information, and data in whatever form in which it is expressed or obtained... Throughout this book, we use the terms knowledge commons and information commons

---

6 This comment is particularly relevant to the work of James Boyle. As has been shown, neither Lessig nor Benkler equates commons with the public domain. In fact, they avoid using this notion, whose outlines are uncertain.

interchangeably" (Hess and Ostrom 2007, p. 7)[7]. In the same way, the notion of knowledge covers the scientific field as well as the artistic field: "Knowledge as employed in this book refers to all types of understanding gained through experience or study, whether indigenous, scientific, scholarly, or otherwise nonacademic. It also includes creative works, such as music and the visual and theatrical arts" (Hess and Ostrom 2007, p. 8).

Information technologies have challenged the usual modes of production, circulation and consumption of knowledge. They have rendered obsolete many of the norms, rules and laws that prevailed until then. New institutional dynamics have gradually emerged, but still open enough to make any prediction about the information ecosystem uncertain. What is certain for Hess and Ostrom, on the other hand, is the need to try to build a barrier against a creeping dynamic of privatization by defending the idea that knowledge, in the digital ecosystem, can be eligible for common status.

If, however, we want to go beyond the rhetorical argument and anchor it in a strong theoretical foundation, they argued, it is necessary to study the properties of the digital ecosystem, which is in many respects more complex, with much fuzzier boundaries, than that of a natural resource. Moreover, an information commons does not share the same economic properties and is therefore not subject to the same social dilemmas as a natural commons. The latter is rival and non-exclusive, which makes it subject to a major social dilemma, identified by the risk of overexploitation and thus the eventual disappearance of the resource. This is what makes it eligible, more than any other natural resource, to be managed as a common pool.

Conversely, an information resource has *a priori* the properties of what the economic approach defines as a public good: it is non-rivalrous (by consuming it, I do not diminish the quantity of the resource available to others) and non-excludable (it is difficult to exclude others from consuming the

---

7 However, they specify a conceptual point of view that these notions (knowledge, information and data) are not identical: "Our thinking is in line with that of Davenport and Prusak (1998, 6), who write that 'knowledge derives from information as information derives from data'. Machlup (1983, 641) introduced this division of data-information-knowledge, with data being raw bits of information, information being organized data in context, and knowledge being the assimilation of the information and understanding of how to use it" (Hess and Ostrom 2007, p. 8). But what justifies considering all these dimensions together in an approach to the commons is their intangible nature which subjects them to the same properties of non-rivalry and non-exclusivity.

information resource). In the case of informational CPR, the dilemma of overexploitation disappears. Thus, for Hess and Ostrom, other social dilemmas may arise depending on the type of CPR: "Typical threats to the commons of knowledge are commodification or enclosure, pollution and degradation, and unsustainability" (Hess and Ostrom 2007, p. 5). The risks of privatization and enclosure primarily concern the domain of knowledge and ideas, whereas the risks of pollution and degradation concern the Internet as an information CPR.

However, for Ostrom, this schema is too simplistic to account for the complexity of knowledge as a common. First of all, the property of non-exclusivity depends on the nature of the CPR that gives it "body", because it is indeed it, in the last resort, that is assimilated to a commons and not the information or knowledge itself.

Let us recall that in the Ostromian schema, it is natural CPRs that have, more than other resources, a predisposition to becoming commons and not directly the resource units that compose them. The latter take the form of facilities that are themselves composed of a set of resource units (a flow) that they generate and that can be subtracted by individuals for specific uses (consumption, sale). Informational CPRs are identified in the form of physical structures that store and make available *information units*.

Figure 2.1 shows the accompanying diagram.

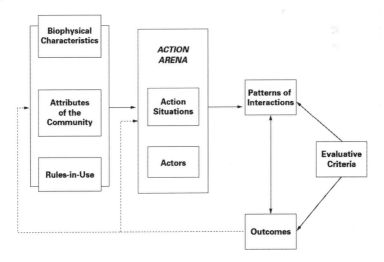

**Figure 2.1.** *Institutional analysis and operating framework*
*(source: Hess and Ostrom 2003, p. 129)*

Under this banner, Hess and Ostrom classified two categories of entities: libraries and archive directories, both physical and digital, and the Internet, an information infrastructure *par excellence* as well as based on physical layers. These two categories are the equivalent of natural resource systems such as forests and pastures. The informational units of these informational CPRs take the form of knowledge, information or data, all identifiable by their intangible content. However, these informational units, defined as ideas, are physically locatable by an *artifact* that constitutes a discrete and observable representation of them: books, scientific articles, a web page, a database or a digital file. The digital ecosystem causes the artifact encapsulating knowledge to lose its physical materiality (in the form of the object "book", for example). It thus becomes more difficult to dissociate the observable and material representation of knowledge from the knowledge itself, the latter taking the form of a digital file circulating in this new space. Moreover, the properties of this new knowledge artifact radically transform the conditions of production and diffusion of knowledge itself. Thus, a common information resource, that is, a common pool resource (CPR), comprises three intrinsically linked dimensions, not two: the artifact, the ideas and the storage space that makes them available. Any institutional analysis of an information CPR must therefore start from such a schema.

## 2.2.2. *Digital libraries as information CPRs*

As Berkman Center researchers Hess and Ostrom agreed, in the face of the threats of enclosures that knowledge faces today, it is essential to think about new institutional arrangements to support its development. They focused their attention primarily on digital libraries[8] in the field of scientific communication, most probably because of Hess' interest, as a university librarian, in the changes and new challenges in this field.

Libraries have characteristics that make them eligible for what they have defined as informational CPRs. Whether physical or digital, they provide access to artifacts in the form of books or scientific journals, where

---

8 According to Lyna Da Silva (2013), a digital library can be defined as a set of digital resources and associated technologies for creating, locating and using information; libraries include data, metadata and links (metadata) to other resources. Beyond their diversity, libraries are a socio-technical innovation in the sense that they constitute a collection of resources sharing the same types of encoding and dissemination, the use of metadata as *resource keys,* and the textual expression of this metadata, regardless of the format of the resource.

knowledge units are stored. They are addressed to a determined community (students, teachers) which has usage rights on these artifacts. Like natural CPRs, these information CPRs are not intrinsically defined by a property regime. Even if digital libraries are eligible to become commons, only the examination of the different property rules, in terms of use and management between the different stakeholders, allows us to determine their exact nature and their robustness.

Already in the case of physical libraries, the governance of these information CPRs is based on ownership rules that resemble a bundle of rights. On the one hand, the user community has rights of use and appropriation over informational artifacts (such as the right to borrow, for example). These are operational rules in the sense of Ostrom. The collective rules in terms of management, exclusion and alienation are defined and held by the governing bodies of these CPRs, which do not, in this sense, have exclusive ownership over the information artifacts. The user community generally does not have such collective rights. This is an important difference from the natural commons studied by Ostrom, whose originality lies precisely in the fact that it is based on rules that have emerged from the commoners themselves on the basis of learning and interpersonal communication. There is a third category of "exogenous" rules defined by the law that apply to informational CPRs, such as the principle of fair use or copyright exceptions, which define certain uses of information artifacts. Finally, as physical storage space, they can only contain a limited number of artifacts. The costs of exclusion in the event of non-compliance with these rules are low compared to other information CPRs. However, these structures are subject to the risk of deterioration if investments are not made in terms of their maintenance.

Any institutional approach that aims to characterize a digital library or archive as a knowledge commons must address the following conundrum: how do individuals create communities, make decisions and build rules in order to preserve and enrich information resources? The first step will be to study the rules, formal and informal, that jointly define the bundle of rights associated with the informational CPR being studied. Thus, starting from the original classification on the bundles of rights for common land, Hess and Ostrom proposed to adapt it to the level of knowledge commons. The first singular feature relates to the fact that the rights of use and appropriation are constrained by the copyright that is imposed on any digital library, whatever its singularity. A researcher who submits an article to a digital archive does not lose his or her copyright. This copyright must also be respected when a user wants to appropriate a resource found in a digital archive. However, we can also imagine archives that choose to circumvent this right by introducing the possibility of using Creative Commons licenses[9] that delimit a different space of use and appropriation. In any case, the rule of open access for all is what is common to all digital archives. This usage rule differentiates them from physical libraries, which grant access rights according to the user's attributes.

Hess and Ostrom proposed to divide the institutional rules defining the set of rights attached to a digital library or archive into seven (rather than five) categories (Hess and Ostrom 2007, p. 52):

– access: the right to enter a defined physical area and enjoy nonsubtractive benefits;

– contribution: the right to contribute to the content;

– extraction: the right to obtain resource units or products of a resource system;

– removal: the right to remove one's artifacts from the resource;

– management/participation: the right to regulate internal use patterns and transform the resource by making improvements;

– exclusion: the right to determine who will have access, contribution, extraction, and removal rights and how those rights may be transferred;

– alienation: the right to sell or lease extraction.

The two new categories are the right to contribute to existing content and the right to delete content, both of which refer to frequent uses on these archives.

---

9 "To provide an alternative to the brittle confines of copyright law, a group of legal scholars developed the Creative Commons in 2002. This service uses 'private rights to create public goods ... a single goal unites Creative Commons' current and future projects: to build a layer of reasonable, flexible copyright in the face of increasingly restrictive default rules'. This collective-action initiative is a case of changing operational rules in order to adapt to evolving technologies and new forms of restrictions" (Hess and Ostrom 2007, p. 52).

## 2.2.3. *Institutional analysis and development framework (IAD)*

Beyond the updating of the components of bundles of rights in the case of information CPRs, the contribution of Hess and Ostrom is also and especially at the level of their IAD methodology (Institutional Analysis and Development framework) that they propose to transpose as a relevant model of analysis in the field of knowledge commons. Digital library projects present a strong heterogeneity in terms of rules of use, modes of governance and financing. They may have been created by and for a community of researchers, by institutional actors (libraries or universities), or by commercial actors (open journals). The institutional rules relating to usage rights (on the submission of articles, for example) are also not uniform. Governance also takes multiple forms. Some digital archive projects have failed, while others have been masterfully successful.

In their book on knowledge commons, Hess and Ostrom defend the idea of applying the IAD methodology initially planned as a grid for analyzing situations of common management of land resources to knowledge commons:

> This framework seems well suited for analysis of resources where new technologies are developing at an extremely rapid pace. New information technologies have redefined knowledge communities; have juggled the traditional world of information users and information providers; have made obsolete many of the existing norms, rules, and laws; and have led to unpredicted outcomes. Institutional change is occurring at every level of the knowledge commons (Hess and Ostrom 2007, p. 43).

The IAD methodology thus constitutes an analytical grid that makes it possible to identify, among the design principles that emerge from the various digital library projects, the factors that can explain their relative robustness: "We expect that the framework will evolve to better fit with the unique attributes of the production and use of a knowledge commons. Over time, it will be possible to extract design principles for robust, long-enduring knowledge commons" (Hess and Ostrom 2007, p. 68). Their contribution is therefore of a methodological nature. As Sene and Hollard point out: "The IAD model should undoubtedly be understood as an operational framework, a reading grid, serving as a basis for an assessment of common management problems. In other words, it is not intended to propose a theoretical model, nor a simplified description of what happens in practice in the field" (Sene and Hollard 2010, p. 449, author's translation).

More concretely, the IAD is based on three clusters of variables, each of which brings together factors likely to influence the institutional design and the pattern of interaction of the common resources under study:

1) the characteristics influencing the decision-making of the common stakeholders: the biophysical properties, the attributes of the community and the rules (formal and non-formal) of the CPR under study;

2) the arena of action: the decisions of common stakeholders in a context of repeated interaction;

3) the results produced by these actions associated with evaluation criteria.

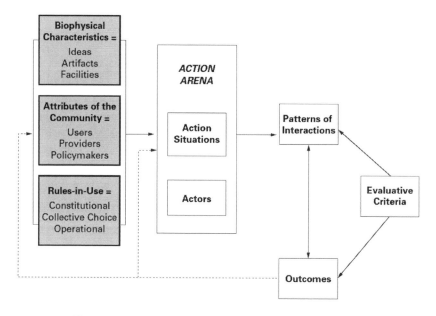

**Figure 2.2.** *Biophysical and institutional characteristics, community attributes (source: Hess and Ostrom 2007, p. 46)*

Figure 2.2 summarizes this analysis methodology. This reading grid can be used in different ways depending on the objective assigned to the study of an informational CPR as a commons.

If we are trying to account for the specific nature of an information CPR, it is necessary to highlight the various factors that directly influence it: the characteristics of the information infrastructure, the specificity of the community in terms of the types of users (knowing that it has much broader and more diffuse outlines than in a natural CPR) and the way in which the different "members" are engaged in the governance of the CPR, and finally, the rules that structure the organization of an open archive (grouping together the rules for the filing and use of the archive such as the rules of ownership that prevail over the information units and the CPR).

In order to understand how the interactional dynamics of different members of the community operate, it is necessary to focus on the arena of action, the strategies of the actors and the nature of their incentives in a given interactional context, which may not be similar, for example, the case of a disciplinary archive or the case of an institutional archive (resulting from the will of an institution such as a university).

Finally, we may also want to evaluate the interaction models produced by an open digital archive. The way the different actors interact strongly affects the success or failure of the resource. Two types of social dilemmas may arise. On the side of the "producers" of the information units, they may not comply with the rules and may not deposit their scientific communications on the archive. On the side of the end-users, we can also see a phenomenon of under-use of the information commons. In terms of results, we can highlight the negative results with first of all the phenomena of enclosure on open resources, as well as the positive results such as the creation of new information commons that produce better access to information, thus contributing to the progress of scientific knowledge.

Hess and Ostrom's contribution has legitimized the creation of a new research program on this issue by providing a solid methodological basis that everyone can use to study the nature and robustness of digital archives in different contexts. In other words, the challenge is to highlight, through comparative empirical studies in the field of open digital archives, what are the "design principles" that can explain the rise of some and the decline of others[10].

---

10 Let us mention that such a research program is currently underway at the University of Ottawa (School of Information Studies), initiated by Professor Heather Morrison: "The

## 2.3. Open access platforms as scientific commons?

Considering digital libraries as scientific knowledge commons (SKC) is a novel theory which, for their instigators, aims to enrich a pre-existing large-scale reflection on the transformation of the contemporary scientific publication and communication regime embodied in particular by the emergence of the open access (OA) movement. The question that we wish to explore here is to evaluate the possible avenues of enrichment that might result from a rapprochement between the approach to knowledge commons and the reflections led by the advocates of open access. This apparent and seductive proximity between open access and commons only makes sense if it goes beyond the space of the common struggle in the activist space and opens the way to a reconfiguration of the ecosystem of scientific communication by offering the opportunity for new, creative and original editorial modes to develop and to hinder any attempt to encapsulate scientific knowledge. This is the path that we choose to follow here[11] in the footsteps left by Hess and Ostrom by mobilizing their institutional methodology, the IAD, which will allow us to qualify the specific nature of OA platforms as common knowledge potentials (which we choose to qualify as scientific to distinguish them from other existing and already identified types). Two issues will be successively addressed, offering an insight into the eligibility of OA platforms as common knowledge: the nature of the bundles of rights attached to OA platforms, then the conditions for their enrichment and sustainability.

---

purpose of the Sustaining the Knowledge Commons research program (and blog) is to advance our knowledge of how to build and sustain a global knowledge commons. It is to focus on the relationship between two basic concepts, 'open access' and 'the commons'. There is an intuitive complementarity between the two concepts that might be best understood as an outcome of recent historical developments". For more information, see: https://sustainingknowledgecommons.org/about/.

11 To our knowledge, the only reflection aimed at applying Ostrom's commons approach in the context of open science is the one delivered by Heather Morrison: Morrison, H. (2019). Open access versus the commons, or steps towards developing commons to sustain open access [Online]. *Sustaining the Knowledge Commons/Soutenir les Savoirs Communs*. Available at: https://sustainingknowledgecommons.org/2019/04/23/open-access-versus-the-commons-or-steps-towards-developing-commons-to-sustain-open-access/. She has made a contrary choice, which is to start from Ostrom's design principles and not from the IAD methodology.

### 2.3.1. *Open access: a major transformation of the editorial ecosystem*

In the scientific field, the advent of digital technology has offered an unprecedented opportunity for new ways of producing, sharing and circulating knowledge more widely and rapidly for the research community. Faced with these new opportunities, from the beginning of the 1990s, new editorial modes have gradually developed, which have resulted in a redistribution of the actors' logics that had prevailed for many decades[12]. The emergence of the Web has been an opportunity for researchers in particular to explore new avenues of academic communication that have materialized in various forms with the creation of open archive[13] platforms, as well as with the creation of scientific journals developing outside the scope of traditional publishers. All of them are characterized by an open access to scientific documents hosted without tariff barriers. This is what some later chose to qualify under the terminology of open access. At the same time, the publishers who controlled the editorial value chain have also taken advantage of these technological changes to implement new strategies that have particularly disrupted contractual relations with libraries, creating a movement of protest on their part and simultaneous support for the various OA initiatives.

Let us return to these different dynamics from a chronological point of view.

At the beginning of the 1990s, new experiments in the construction of digital platforms for self-archiving scientific publications were launched. The idea of grouping together within a centralized automatic repository of as yet unpublished versions of scientific articles was realized for the first time in 1991, by physicist Paul Ginsparg with the creation of arXiv.org. For several years, the community to which he belonged had been in the habit of sending

---

12 See Guédon (2017) for an enlightening overview on the history of what he calls the scientific conversation.

13 Open archives are defined by collections or databases of articles available online made up from the authors themselves who make a voluntary deposit approach. The deposited resources include all the productions that structure the research activity and not only articles already published elsewhere in journals. We can deposit preprint articles, but also conference papers, data sets, theses, dissertations, etc.

pre-printed[14] articles by e-mail; but with the birth of the Web, the possibility of having a centralized server offered a new opportunity to democratize the exchange of information and to accelerate access [15] to it. The initial objective was not to substitute this self-archiving process for the traditional publishing process, but rather to offer a device facilitating, upstream, the sharing of knowledge and its discussion between actors of the same scientific community. It offered a novel way for physics researchers to be able to quickly consult recent publications without waiting for the traditional validation procedure by proofreading, which is much slower. The innovation introduced by this digital archive is, as Jean-Max Noyer argues, "to have opened up the possibility of creating, within the framework of new encyclopedic forms, high-performance editorial devices constructed as 'associated environments' (in Simondon's sense), meeting the requirements of current modes of construction of scientific knowledge" (Noyer 2010, p. 63, author's translation). The notion of publication disappears because the articles deposited become, by definition, moving, as other versions can be deposited with modifications within a very short period of time following comments made by third parties.

An open archive does not have a reading committee, in the strict sense of the term, unlike an open journal, even though all the articles sent in undergo some form of technical control ensuring their universal accessibility and minimal human control to verify that the article belongs to the discipline, but without scientific examination. However, it would be erroneous to conclude that there is no form of evaluation. Other forms of evaluation are proposed, such as the hearing of articles filed. This is an issue that is the subject of much debate and which classical publishers have seized upon to discredit these self-archiving platforms. The success was greater than expected. In six months, the number of requests for submissions had multiplied by 4. This

---

14 In the field of scientific publication, a prepublication (also called author manuscript and very frequently preprint) is a version of a scientific article that precedes its acceptance by the editorial board of a scientific journal.

15 "The original bulletin board was engineered to level the research playing field. It is hard to imagine now, but considerable time and effort was once spent printing, photocopying and posting preprints to a privileged few friends and colleagues, before publication in formal journals. The idea of a central repository was to allow any researcher worldwide with network access to submit and read full-text articles, giving equal entry to everyone from graduate students up. (The early Internet was an academic playground — the general public didn't start coming online until a few years later.)" Extract from Ginsparg, P. (2011). ArXiv at 20. *Nature*, 476, 145–147. Available at: www.nature.com/articles/476145a#citeas.

archive has also broadened its scope of publication, originally restricted to the field of high-energy physics, to cover all areas of physics. Thus, in 2011, 20 years after its creation, arXiv had accumulated 700,000 submissions and received more than 75,000 new ones each year, for a total of 1 million full text downloaded[16].

The success of this pioneering initiative has undeniably encouraged the subsequent creation of scientific open archives along the lines of arXiv.org and more broadly within the global open archive movement[17]. In France, the physicist Franck Laloë is credited with having worked on the internationalization and consolidation of arXiv in close collaboration with Paul Ginsparg[18]; this platform subsequently gave birth to the open archive platform in France. He promoted the creation of the CCSD (*Centre pour la communication scientifique directe*), an entity attached to the CNRS which took charge of the creation of the first open archive in France, under the name of HAL (*Hyper articles en ligne*). HAL presents itself as an archive of scientific documents, published or not, and theses produced in the context of scientific research and higher education, public and private, French or foreign. Unlike arXiv, it is not reserved for some scientific disciplines. Very quickly, its vocation is to centralize all research produced in all disciplines[19]. Shortly after its launch, there was a desire to enhance the value of documents collected in institutional collections in particular. This archive thus went

---

16 These data are those provided by Paul Ginsparg in the article "It was twenty years ago today...", available at: https://arxiv.org/abs/1108.2700v2.

17 Among the best known are PubMed Central, an open archive of the National Institutes of Health (www.ncbi.nlm.nih.gov/pmc/) in the biomedical and life sciences, and Research Papers in Economics (http://repec.org/), an open archive in the fields of economics.

18 "From 1994–1995, I was struck by the growing impact of the Los Alamos base. But the functioning of this base posed obvious problems: everything depended on one person, Paul Ginsparg, and one country, which introduced an imbalance that was not desirable if a new universal level of communication was being created. It was necessary for this base to acquire an international mode of operation. Paul Ginsparg and I set ourselves this objective. I convinced him after lengthy discussions that it was necessary to build a system that eluded him: 'Okay, but it's up to you to do something about it. concrete and not just ideas,' he once concluded – that was the starting point" (Barberousse et al. 2003, p. 180, author's translation).

19 "While physics and mathematics are for the moment the most involved disciplines, other CNRS departments are also interested, especially the HSS department (Humanities and Social Sciences). On the other hand, chemists, for example, show little interest, for reasons that are surely excellent and have to do with the intellectual habits in their discipline. It is clear that it will never be a completely universal tool, if only because of the specific problems linked to military or industrial applications" (Barberousse et al. 2003, p. 180, author's translation).

beyond the strict spectrum of communication between researchers. Today, it is the most important database for francophone research with more than 1 million publications. Since 2013, HAL has had a singular role within the ecosystem of open archives in France, as it has been designated as "the shared national infrastructure hosting institutional archives or to which other institutional archives are strongly invited to contribute their content"[20].

Alongside HAL, in France, some open disciplinary or thematic archives also saw the light of day, like the @rchivSic[21] site, created in 2002 by Ghislaine Chartron and Jean-Max Noyer, joined later by Gabriel Gallezot, in the field of information and communication sciences. For its co-founders, the creation of such a digital archive within the ICS community did not answer, as for the physicists' one, to a problem of sharing articles in pre-print version, which was not a usual practice. For their instigators, the goal was elsewhere. As a new editorial mode, this self-archiving platform offered the opportunity to develop new ways of representing collective intelligence, a new political economy of knowledge based on new socio-cognitive models and intellectual practices[22]. We will return to this because it is an essential dimension that is rarely emphasized and yet constitutes the major interest from the point of view of the anthropology of sciences.

For their part, commercial publishing houses have not remained inert in the face of the new conditions of production, distribution and circulation of knowledge offered by the digital ecosystem. They have taken the opportunity to review in depth the nature of transactional relationships with libraries. With the emergence of the Web, the logic of exchanging property rights on material objects in markets has given way to a logic of access to content on the network. Libraries have quickly appropriated this new *modus operandi* not by giving more rights of use to readers, but, on the contrary, by restricting them. In particular, they have taken advantage of this change in

---

20 This decision was taken within the framework of the signature at the Académie des sciences on April 2, 2013 of the Partnership Agreement in favor of the open archives and the mutualized platform HAL; quoted page 3 of the report of the CNRS and the Direction de l'information scientifique et technique , "L'*open access* à moyen terme: une feuille de route pour HAL", September 2014.

21 https://archivesic.ccsd.cnrs.fr/.

22 In issue 1 of the journal *Solaris*, Jean-Max Noyer already explains what constitutes the fundamental issues related to the process of digitization of the sign in the scientific field and what he calls the emerging digital plasticity. See: http://gabriel.gallezot.free.fr/Solaris/d01/1noyer1.html.

the ecosystem to increase their control over the content they offer. In doing so, they have gone beyond what print and copyright had established by introducing a kind of silent revolution in the transactional legal framework (Guédon 2017). Thus, the principle of exhaustion of rights that prevailed in the universe of physical journals (and defined by the legal framework) disappeared in favor of a contractual logic, decided unilaterally by publishing houses, concerning the control of copies of digital journals. Large publishers have also implemented a marketing strategy known as the "big deal", which requires libraries to buy access to their entire catalog (and not just a selection of journals as they used to do), generating a significant additional cost that leads them to abandon other smaller publishers or to reduce the purchase of paper books. As Peter Suber notes, "According to the American Association of Research Libraries, between the mid-1980s and the mid-2000s, the price of paid journals rose 2.5 times faster than the cost of living" (Suber 2016, p. 45, author's translation). This price increase very often exceeded the library budget. The result has been a crisis in access to paid access peer-reviewed journals that affects even the most affluent universities.

Faced with a profound shift in their information ecosystem and the predatory strategies of scientific publishers, some of the academic library communities have been prompted to react to protect public access to these resources. Nancy Kranich, former President of the American University Libraries (2000–2002), echoes this sentiment in Hess and Ostrom's book: "In the face of these enclosures, librarians along with their colleagues in the scholarly community have struggled to protect access to critical research resources, balance the rights of users and creators, preserve the public domain, and protect public access for all in the digital age" (Hess and Ostrom 2007, p. 92). Among the initiatives that have contributed to introducing new approaches to manage and disseminate their collective information resources, Kranich cites two flagship projects. Firstly, the SPARC (Scholarly Publishing and Academic Resources Coalition) project, founded in 1998 as an international alliance of research libraries, universities and other organizations (300 members in total). It was conceived as a constructive response to the dysfunctions of the academic communication market and aims to support digital archive projects and the promotion of initiatives towards the democratization of access to scientific information. Secondly, the Open Archive Initiative (OAI), created in 1999 by the community of librarians, is a project that aims to promote the exchange and development of digital archives through the implementation of an architecture and technical tools that allow interoperability between different archive servers. The concrete result of this initiative has been the invention of an OAI-PMH (Open Archives Initiative Protocol for Metadata Harvesting) protocol allowing interoperability between scientific archive servers. Subsequently, the *Eprints* software was invented to generate archives according to OAI standards. Thus, all archives using this protocol can be queried as if they were one.

Finally, a third dynamic has contributed to the reconfiguration of the scientific editorial ecosystem, with the emergence of 100% online journals that have freed themselves from the traditional intermediation of publishers, marked by a desire to regain control of the scientific knowledge production chain (Guedon 2017). This is what we call open publishing. It has the same operating principle as traditional journals, with one difference: it is free of charge for the user. Accessibility to the open edition is promoted by the existence of dedicated platforms such as the Directory of Open Access Journals (DOAJ), which lists more than 10,000 open access scientific journals from more than 130 countries. In France, the creation of the Revues.org platform by researcher Marin Davos in 1999 was a pioneering initiative[23]. To date, it offers more than 500 journals in the field of humanities and social sciences, 95% of which are freely accessible content. There is no longer any barrier to entry imposed by the publisher on the reader.

Along the way, all these initiatives from both researchers and librarians have gradually led to the constitution of a true intellectual movement in defense of open access. This has found fundamental political support through the active support of the Open Society Institute by the Hungarian-born billionaire and financier George Soros. Soros created this institute in 1979 with the aim of promoting democratic governance and human rights in particular. It is closely associated with the Budapest Open Access Initiative (BOAI)[24] organized in 2002, which marked the institutionalization of open access as a movement for the free dissemination of scientific production on the Web.

---

23 Since 2016, Revues.org has been integrated into the OpenEdition portal, which is a complete infrastructure for electronic publishing at the service of scientific information in the humanities and social sciences. The OpenEdition portal is international in scope and includes four platforms for publishing and information in the humanities and social sciences: OpenEdition Journals (journals), OpenEdition Books (collections of books), Hypotheses (research notebooks) and Calenda (announcements of international academic events).

24 www.budapestopenaccessinitiative.org/read.

At this conference, the Open Society Institute pledged to provide support and initial funding to expand and promote institutional self-archiving, to launch new open access journals, and to help make the open access journal system economically self-sufficient. While the commitment and resources of this institute are substantial, this initiative notes that it is in great need of the efforts and resources that could be provided by other organizations[25].

Among the signatories of this initiative are intellectual figures who have since played a fundamental role in the defense of open access, such as Steven Harnard, Peter Suber and Jean Claude Guédon[26]. Here is the definition of open access:

> By "open access" to this literature, we mean its free availability on the public internet, permitting any users to read, download, copy, distribute, print, search, or link to the full texts of these articles, crawl them for indexing, pass them as data to software, or use them for any other lawful purpose, without financial, legal, or technical barriers other than those inseparable from gaining access to the internet itself. The only constraint on reproduction and distribution, and the only role for copyright in this domain, should be to give authors control over the integrity of their work and the right to be properly acknowledged and cited[27].

This conference also formalized in a very explicit way the two possible ways of developing open access: self-archiving and open access journals. The first is still called the green way and the second the golden way.

In the continuity of this pioneering initiative, several other demonstrations over the last decade have supported this OA movement. However, while OA was originally a bottom-up movement, originating in the academic and library communities, more recently, it has been the object of sustained attention from political decision-makers, first at the European level, then relayed at the national level. The latter will progressively play an

---

25 *Idem.*

26 Peter Suber is Professor of Philosophy at Earlham College and a member of The Free Online Scholarship Newsletter; Stevan Harnad is Professor of Cognitive Science at the University of Southampton and the Université du Québec à Montréal; Jean-Claude Guédon teaches at the Université de Montréal.

27 www.budapestopenaccessinitiative.org/read.

increasingly prescriptive role in what the European Union now calls open science[28] (Vanholsbeeck 2017)[29].

However, this late interest in open science reveals for Vanholsbeeck (2017) ambivalent, even paradoxical objectives. On the one hand, it marks an institutional willingness to open up research results beyond the restricted circle of scientists, in order to foster its appropriation by both economic actors and citizens themselves. On the other hand, it also reflects a desire to increase the managerialization of research. From 2016, the European Union wished to see a rapid shift in the editorial ecosystem towards OA being the default principle from 2020. In 2018, Plan S was launched, a new initiative from the European Commission supported by the European Research Council and research funding agencies in 12 European countries to promote open access scientific publishing. At the national level, in France, the Ministry of Education and Research launched a National Plan for Open Science[30] announced by Frédérique Vidal on July 4, 2018, making OA mandatory for publications and for data from project-funded research. As pointed out by Chartron and Schöpfel (2017), the transformation of the editorial research system towards an open ecosystem without access barriers now appears irreversible, as a political objective as well as a strategy for higher education and research institutions and organizations, or as a business strategy, as far as the information and publishing industry is concerned.

This overview of the genesis of open access reveals the intersecting and heterogeneous dynamics that constitute the OA movement. It also reveals a fundamental underlying question, that of the political knowledge economy, which is being radically transformed under the effect of these intersecting dynamics.

Indeed, the institutional will to question the oligopoly of scientific publishers leaves open the question of new institutional arrangements on which to base the scientific publishing ecosystem: what governance? What

---

28 While OA focuses on the conditions of access, open science is about how researchers work, cooperate, interact and share resources and disseminate their scientific outputs.

29 OA was thus the first area to be subject to European requirements in the field of what the Commission first called "Science 2.0", and then, from 2014 – following a public consultation – "Open Science" (Commission 2015). From 2012, the Commission put OA at the heart of a communication (Commission 2012b) and a recommendation (Commission 2012c) dedicated to access and preservation of scientific information.

30 www.enseignementsup-recherche.gouv.fr/pid39205/science-ouverte.html.

socio-economic models? Considering OA platforms as potential new forms of knowledge commons[31], as Hess and Ostrom suggested, is a relevant avenue of study to shed light on these transformations. Here, we take this path of investigation by relying, as they prescribe, on the IAD methodology previously highlighted as a reading grid.

### 2.3.2. *Open access platforms: which bundles of user rights?*

The first step in Ostrom's IAD methodology is to study the biophysical properties, community attributes and rules (formal and informal) of the informational CPR being analyzed. Its purpose is to specify the singular nature of the knowledge commons. We will begin, firstly, by highlighting the rules that regulate the conditions of access, appropriation, contribution and the collective rules of governance of OA platforms as potential informational CPRs.

Let us return to the Budapest Open Access Initiative. This initiative insists primarily on the definition of open access in terms of user rights. This can be seen as a point of convergence with Hess and Ostrom's knowledge commons approach, which defines open access in terms of bundles of rights. This proximity therefore deserves to be studied in more detail, especially since we will show that there is not one but several possible interpretations of the bundle of rights associated with the OA approach. Indeed, Peter Suber (2016) states that, following the BOAI, two other public declarations[32] structuring the OA movement have taken place; they are important because they adopt a more restrictive vision of user rights. These divergences of

---

31 Hess and Ostrom have associated their book on knowledge commons with leading figures in the United States who are committed to open access to scientific knowledge in the digital ecosystem, including Nancy Kranich, past president of the American Association Library Association, and Peter Suber, a professor of philosophy and an early advocate of open access. For Kranich, seeing digital science archives as new forms of knowledge commons is a strategy for paving the way for a new paradigm for the creation and dissemination of scholarly communication, fostering a new model of information sharing and the emergence of new forms of creativity. For his part, Peter Suber (2007) argues for a rapprochement between the open access movement and the commons movement, since both take a similar view of the rights of users to use knowledge made available in archives without asking permission from rights holders.

32 The Bethesda Declaration on Open Access Publishing, signed on June 20, 2003, and the Berlin Declaration on Open Access to Knowledge in the Sciences and Humanities, signed on October 22, 2003.

interpretation did not lead to a consensus and, even today, these divergences persist. The scientific content platforms in OA present disparities that cover the initial oppositions between these different initial conceptions. These dissensions are important because they testify to the existence of several categories of common scientific knowledge. Highlighting them is an indispensable first step in order to be able to then assess the conditions of enrichment and sustainability of each of these categories.

In all cases, all platforms in OA, green or golden, offer the user the possibility to have a right of access free of charge and without prior permission (from the author or the platform) to the full content of the proposed scientific documentary resources. The abolition of the tariff barrier is what constitutes the distinctive and uniform feature of OA platforms. As mentioned by the supporters of OA, the authors of scientific articles are not traditionally remunerated by publishers; therefore, they have nothing to lose by making their articles available for OA as long as they retain their authorship rights (each applicant retains their intellectual property rights, including the right to be properly cited and recognized as the author of a document). In addition, they can benefit from a much larger audience.

Beyond this right to read the texts in full, there are "appropriation" rights that can be granted by an OA platform. In this case, several possibilities exist, which depend on the choices made with regard to copyright by the OA platforms as well as on the type of documents filed.

We will start with self-archiving platforms. Two main categories of documents can be submitted: preprints from unpublished research work and postprints from scientific articles previously published in a journal. In the case of preprints, the author is *de facto* protected by copyright, but the self-archiving platform may have a policy allowing the use of free licenses for authors.

This is the case of the open archive platform HAL, in France, which offers depositors several licensing options: the all-rights reserved license (classic copyright), the full range of Creative Commons licenses, the Etalab license and the public domain. Here is what is specified on HAL on this subject: "Creative Commons licenses can be used in a complementary way to copyright (sharing, reproduction, evolution, etc.). The CCSD advises to inform this data in order to better inform the reader of the rights granted by the author as well as to allow a better use of the data during an automatic exploitation"[33].

By introducing the possibility of an open license for applicants, it can be said that HAL offers a range of potential uses beyond the standard of open access to read-only text and the right to copy short citations for certain articles. The introduction of open licenses is not neutral, because they give the user greater freedoms that increase the possibilities of sharing and dissemination of scientific content: citing long passages from a given article; distributing this article in its entirety to students or colleagues; distributing modified or augmented versions of this article, for example, versions with semantic tags; converting this article for reading in new formats or on new media following technological advances; including this article in a database; using this article for indexing, data mining or any other type of processing, etc. (Suber 2018, p. 84).

Figure 2.3 results from a collective work of students studying for the DASI (Data analytics and information strategy) master's degree at the UFR Ingemédia de Toulon[34]. It gives the relative proportion of articles deposited on HAL which use an open license. It allows us to note that to date, a very tiny number of the deposits on HAL are put in CC license. Why is this so? The question remains still open at this stage of our study. Several hypotheses can be put forward and, among them, are a lack of information about these licenses, the rights that they confer and the interest, from a scientific point of view, of using them.

However, not all self-archiving platforms offer the same choice in terms of licenses. For example, @rchivSic does not offer the licensing option open to their depositors. The rights of use are therefore more restricted and there is no sharing possible. This means that the author of a scientific article may therefore require his or her consent when users copy, use, disseminate, transmit and publicly display the work. However, the user can still reuse short excerpts without asking the author's permission in accordance with the exception clause of copyright law.

---

33 https://doc.archives-ouvertes.fr/wp-content/uploads/2018/03/Charte-de-la-mod%C3%A9ration-in-HAL.pdf.

34 These data come from a work entitled "Open access: les plateformes d'autoarchivage", carried out by Guillyan Chapput, Valeria Sodoma, Alaric Tabaries and Habib Zemni as part of my Master's DASI course, December 2019.

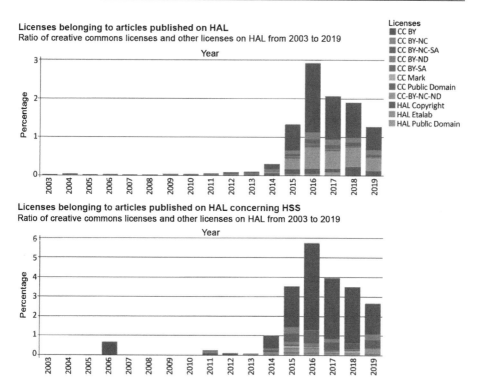

**Figure 2.3.** *Licenses used on the HAL platform (source: HAL). For a color version of this figure, see www.iste.co.uk/pelissier/commons.zip*

Through these two examples of HAL and @rchivSic, we could highlight two categories of bundles of rights which are not identical structuring the uses of open archives. Can we deduce from this that only open archives authorizing the use of open licenses constitute authentic scientific commons while the other category constitutes a non-legitimate version? The question must be asked, but at this stage, this argument does not seem to us to be a sufficient condition. It is also necessary to examine the collective rules and the functioning of governance.

In the case of articles already published in a classic journal (preprint), the problem arises differently. First of all, open archives have the obligation to ask the custodians of the articles to take the steps themselves to obtain the right to deposit their preprint on their platform (the custodians having assigned their copyright to their publisher before publication). A number of traditional publishers give this permission, but a number still refuse it or require an embargo. According to Steven Harnard (2015), about 60% of traditional journals allow their articles to be deposited in open archives, but this permission is not necessarily publicized to researchers[35]. On the other hand, it is very rare for publishers to allow authors to place open licenses on their articles at the same time. Today, in France, the law for a digital republic has clarified the situation by limiting the embargo to six months for an article published in a science journal and to one year for journals in the humanities and social sciences[36]. This law has the merit of reducing the legal complexity that the researcher faces when he chooses to deposit an article already published in an open archive. It is also in line with the European Union's desire to improve access to scientific[37] information. In this case, the bundles of rights attached to the use of postprints in open archives are regulated by formal rules under national law.

For journals under the open publishing regime, the rights of use granted to published articles depend on the journal's policy. The French portal OpenEdition, specializing in the humanities and social sciences, includes 534 journals whose access modalities may confer different usage rights to users. The journals are classified according to three categories of access: open access, which is free access to read the full text in HTML format with the possibility of downloading the article in PDF; open access freemium, which is free access to read the full text in HTML format and with downloading in PDF format possible if the researcher's supervising

---

35 Harnad, S. (2015). Open Access: What, Where, When, How and Why. In *Ethics, Science, Technology, and Engineering: An International Resource*, Britt Holbrook, J., Mitcham, C. (eds). MacMillan, Farmington Hills.

36 This is Article 30: "When a scientific paper resulting from a research activity financed at least half by grants from the State, local authorities or public institutions, by subsidies from national funding agencies or by European Union funds is published in a periodical appearing at least once a year, its author has, even after having granted exclusive rights to a publisher, the right to publish it, the right to make the final version of the manuscript available free of charge in an open format, by digital means, subject to the agreement of the possible co-authors, if the publisher himself makes it available free of charge by digital means or, failing this, after the expiry of a period of time from the date of the first publication. This period is a maximum of six months for a publication in the field of science, technology and medicine and twelve months for a publication in the field of humanities and social sciences". Available at: www.legifrance.gouv.fr/eli/loi/2016/10/7/2016-1321/jo/article_30.

37 On July 17, 2012, the European Commission published a recommendation to Member States to improve access to scientific information and to boost public investment in research. Available at: http://ec.europa.eu/research/science-society/document_library/pdf_06/recommendation-access-andpreservation-scientific-information_en.pdf.

institution subscribes; and mobile barrier, which defines a period during which articles published in a journal issue remain in paid access before moving to free access on the journal's site or its distribution platform. It allows the publisher or the publishing academic society to continue to receive income through per-issue sales or subscriptions for access to recent issues, whether paper or digital. We should add here the possibility of a fourth category of access which does not concern the journals of the OpenEdition portal but which, for all that, has been adopted by a growing number of journals in OA. It is a form of access without a tariff barrier to the reader, but which in fact moves the barrier to the level of the author who will have to pay for his article to be published in this type of journal.

These four possible types of access are in fact distinct economic models. Open access journals are externally funded, either by donation or by government. Open access freemium journals are funded by some libraries that wish to support this initiative by paying for access that confers additional rights in terms of use for their own readers. In the case of mobile barrier, access being temporarily restricted, it is this tariff barrier that thus makes it possible to finance the journal. In the latter case, with the limitation of the embargo period by law, even so-called closed journals must adopt this model of mobile barrier. However, this creates confusion that makes it difficult at first glance to distinguish between journals originally in OA and closed (or so-called hybrid) journals.

Beyond access rights, we must now look at appropriation rights. Open access or open access freemium journals do not adopt the same licensing policy, with some choosing "all rights reserved" licenses, while others opt for Creative Commons (but not very permissive) licenses. Moreover, within journals with identical access rights, the conditions of appropriation are not necessarily the same[38]. At this stage, it would be useful to have a much more comprehensive view in order to be able to better understand the licensing policies of all the journals on the portal and the reasons that led them to

---

38 In the field of information and communication sciences, if we take two journals in the list of those offering open access, the journal *TIC et Société* opts for a CC BY NC ND license, while the journal *Comunição e sociedad* offers the CC BY NC license. Among those proposing open access *freemium,* we find this heterogeneity. The journal *Terminal* offers all its articles with an "all rights reserved" license. Conversely, a journal such as the *Revue française des sciences de l'information et de la communication* (RFSIC) offers its articles under the open license CC BY NC SA, whereas, for example, the journal *Communication* offers them under the open license CC BY NC ND.

make such choices. In the international platform with the largest number of journals in OA to date, the Directory of Open Access Journals, Peter Suber found that the proportion adopting open licenses is still low, but not totally non-existent. Of all open access journals (6,497), only 21.1% use a Creative Commons license: "Most of them use a copyright with all rights reserved that leaves their readers no freedom other than the right to quote" (Suber 2016, p. 83, author's translation). It does not tell us more about the type of licenses used.

What we retain for our present study is that OA platforms do not define the same rights of use concerning the appropriation and reuse that can be made of their content. It has been shown that several of their rules are combined: copyright, the rules enacted by OA platforms, the rules enacted by publishers (green postprint) and finally recent legislation on restricting editorial embargoes (such as the law for a digital republic). In order to make progress in this reflection, it might be wise to categorize these platforms as OA according to the nature of their legal bundles, from the most to the least permissive.

Such a study on the bundles of rights attached to these platforms would be incomplete if it did not pay attention to collective use rights as defined by Hess and Ostrom. These collective rights, which define the perimeter of regulation and management of operating rights (in terms of use), provide valuable information on the modalities and objectives of governance, which constitutes one of the essential dimensions of a common knowledge base. This is an investigation that, here too, must be conducted in depth in a comparative perspective. At this stage, we can only outline a few avenues of research that seem relevant to us, if we cannot provide a demonstrative answer.

For Ostrom, an important dimension structuring the land commons is the existence of a community that plays an important role not only in the emergence process of operating rules, but also in their subsequent regulation. However, in the case of knowledge commons such as open archives, the community here is much more diffuse[39] and more difficult to identify, because access is not restricted to a certain category of actors as in the case of

---

39 "In contrast to the situation with a fishery or groundwater basin, it is much more difficult to grasp who the entire community is that is contributing to, using, and managing a knowledge commons" (Hess and Ostrom 2007, p. 48).

land commons. Moreover, as Hess and Ostrom suggested, it also relies on a heterogeneous set of actors: the platform's users, as well as the providers, those who make the content available as well as the infrastructure, and the policy makers related to self-governing internal communities that can influence the rule-making process (such as members of the FOSS movement or the Open Archive Initiative, etc.). Through this reading grid, we can therefore hypothesize that the heterogeneity of the communities associated with the different OA platforms will have a strong impact on their governance modalities, as well as on the conditions of their enrichment and their sustainability in the longer term. However, these questions can only be elucidated through micro-situational fieldwork, specific case studies, as Hess and Ostrom invite us to do.

In the case of open archives, we mentioned that some emanate from initiatives of researchers within a scientific community (like arXiv) according to a bottom-up logic, while others are the result of an institutional will (like HAL). It can be hypothesized that these differences will have a real impact on the nature of the governance of these platforms and the related collective rights. For example, if we compare the open archives arXiv and HAL, their rules of governance differ in several respects.

Ten years after its creation, the arXiv site was attached to the Cornell University Library. While Cornell University finances about 37% of the operational costs, the site calls on institutions that make significant use of the site to participate in funding. The funding is therefore shared among several partners representing a large part of the user community. Governance is based on a triad in the sense that it is based on three distinct entities, each of which groups together distinct entities with specific functions[40]: Cornell University, which is in charge of the operational management of the platform; an advisory board comprising institutional members such as research institutions, laboratories and foundations that contribute financially to the project, with an advisory role related to the management and development of the standard implementation repository, interoperability and development priorities; a scientific advisory board composed of researchers and scientists from the fields covered by arXiv, whose role is to provide guidance to Cornell University on the issue of intellectual property, the moderation system, criteria and filing standards. Even if the operational and managerial center of this archive is centralized, the different categories of

---

40 https://arxiv.org/about/principles.

users have collective rights in the sense that they participate in the evolution of the operating rules.

HAL is an institutional archive which is defined as a centralized public service offer steered by a public entity, the CCSD, through two distinct entities: the steering committee and the scientific and technical committee (STC). The question of HAL's governance is at the heart of the concerns of its roadmap[41] as already mentioned in the Bauin report (2014). But more generally, it is within the white paper edited by the Scientific and Technical Information Department of the CNRS, entitled "*Une science ouverte dans une République numérique*" (2016); it is stated that there is a need for new governance in the face of the multiplication of platforms (institutional, thematic, deposit, bibliographic, archiving, etc.) and the weakness of their contractual framework (they have more or less restrictive general conditions of use for the use of IST, not always in line with the intellectual property law and publishing contracts between researchers and publishers). This need for governance must respond in particular to the search for a balance between the dual institutional will to promote open access and the strong incentive to file patents: "This requires drawing a dividing line between the common good and protected innovation, between freedom of access and private reservation" (Livre blanc, p. 35, author's translation).

### 2.3.3. *Enrichment and sustainability of the scientific commons*

In order to extend the study of OA platforms as common scientific knowledge, it seems important to us to return to the purpose of such platforms and the objective pursued by their institution. As a preamble, it is useful to recall that if, for land commons, the issue is to determine the

---

41 "It is necessary to make a clear distinction between the STC's function of strategic analysis, advice and decision-making, and the steering committee's function of arbitration and decision-making, in short, exercising supervision. The steering committee should closely resemble a Steering and Monitoring Committee for a CNRS service unit. The STC, for its part, should function as ... a scientific council! The members of the two committees should also be clearly differentiated: among the STC, the specialists, who do not have to represent an institution. In the steering committee, we will find above all the decision-makers, who themselves represent their institution well: they are generally not the same people... It would also be useful to have several committees of users or users by category". In Baouin, S. (2014, author's translation). *L'open access à moyen terme: une feuille de route pour HAL*. Report, CNRS, Direction de l'information scientifique et technique, 9.

institutional arrangements that promote collective action and avoid overexploitation, it is quite different for knowledge commons.

Indeed, as has been said, the social dilemma is not of this nature; the challenge identified for such a commons is to define the design principles that promote its enrichment and sustainability in the face of social dilemmas that need to be identified:

> Potential problems in the use, governance, and sustainability of a commons can be caused by some characteristic human behaviors that lead to social dilemmas such as competition for use, free riding, and overharvesting. Typical threats to knowledge commons are commodification or enclosure, pollution and degradation, and nonsustainability (Hess and Ostrom 2007, p. 5).

But what exactly does this enrichment issue mean in the precise context of open access platforms? What are the conditions guaranteeing such enrichment and their sustainability? The answer to this question is not immediate. Referring to Ostrom's methodology, it is useful to focus on the second part of her schema, which is the arena of action: "Action arenas consist of participants making decisions within a situation affected by the physical, community, and institutional characteristics that will then result in varying patterns of interactions and outcomes" (Hess and Ostrom 2007, p. 53).

### 2.3.3.1. *The enrichment of a scientific commons*

Beyond their specific nature, the enrichment of scientific commons can be defined as the manifestation of a double positive dynamic of the deposit of scientific documents and use of these same documents. This double open cumulative dynamic is essential to foster a dynamic of quality research and economic innovation. This has been the objective of the BOAI from the outset: "Open access for publicly funded research benefits taxpayers and improves the return on investment of citizens in research. Open access produces economic benefits as well as scientific progress"[42]. OA is therefore now at the heart of the conditions fostering scientific progress and the innovation dynamics that it induces. Here, we find Stiglitz's argument (discussed in section 1.5.2). Still from an economic point of view, this

---

42 See BOAJ.

dynamic of opening up scientific knowledge also offers an unprecedented opportunity for the scientific community to regain control of the publishing ecosystem which, in recent decades, has been transformed into an oligopoly dominated by a few major international publishers[43]. The enrichment of these platforms thus constitutes an opportunity for the research community to reappropriate the scientific renewal of their intellectual productions. The latent question is to recognize how the cohabitation between common scientific knowledge and the major publishers is envisaged. At this level, there is no consensus.

From the point of view of the anthropology of science, the question of the enrichment of an SKC can also find an interesting and complementary light to the socio-economic vision presented above. For Jean-Max Noyer (2010, 2015), the new anthropological stratum constituted by the digital ecosystem has paved the way for a bursting encyclopedism, embodied by a vast open domain of internal relations, giving rise to new forms of hypertextual writing and reading, mapping and indexing, and networked digital memories. It thus offers the possibility of developing new digital editorial modes that allow not only a greater visibility of research productions, but also a better consideration of texts at more differentiated stages of their "production–circulation–validation–legitimization" mode. This is a unique opportunity to develop a process-oriented research publishing process, based on a new mode of writing and forms of open editorial objects already envisaged by Ginsparg, the founder of arXiv, in 1996 in the form of an Overlay Journal (Gallezot and Noyer 2010). The documentary question should no longer be thought of as a problem of access, but as a problem of knowledge production and navigation of the movement of knowledge. The problem of access to knowledge distributed on the surface of the planet is no longer the central problem. The main challenge is to take advantage of the progress made in the field of automatic processing of digital traces to set up devices (such as specialized search engines) allowing access to the understanding and exploration of the documentary space and its heterogeneous and open corpora, paving the way to collaborative working modes operating independently of a central instance.

---

43 See the first empirical study that showed such a result: Larivière, V., Haustein, S., Mongeon, P. (2015). The Oligopoly of Academic Publishers in the Digital Era [Online]. *PLoS ONE*, 10(6), e0127502. Available at: https://doi.org/10.1371/journal.pone.0127502.

### 2.3.3.2. *The enrichment conditions of a scientific community*

Such an enrichment dynamic, while desirable, is nonetheless conditioned by a set of factors that promote growth. We have identified two independent categories: the question of individual incentives for researchers to collaborate in the development of OA platforms, and the question of socio-technical choices made by governance in terms of collaboration (one of the ways of enriching the pool of scientific knowledge being that it can be linked together to facilitate its exploration).

Studying the arena of action consists of focusing on the space of individual incentives of different users in terms of contribution to OA platforms: "In analyzing situations, one is particularly concerned with understanding the incentives facing diverse participants. With an institutional repository, many incentives exist for faculty to want to submit their research" (Hess and Ostrom 2007, p. 55). As with the question of bundles of rights, the study of the conditions for enriching the SKC is not *a priori* posed in the same terms for self-archiving platforms and OA journals.

Take the case of self-archiving platforms. Enrichment, understood in the sense of a continuous growth of the flow of scientific document repositories and the use of these same documents, is based on two minimal conditions.

In the first place, this enrichment depends directly on the researcher's incentive to systematically deposit his scientific productions on such platforms (institutional or disciplinary). However, to do so, he must already be aware of their existence and of the modalities of deposit. If, for some platforms such as arXiv, the whole physics community is accustomed to make regular use of them, either by depositing their preprints or by consulting those of others, this is far from being a practice by all researchers, a large part of whom still do not publish in an open archive. This practice of depositing and consulting results from habits specific to a discipline, as we have already mentioned. For a community for whom the publication of preprints in an open archive does not constitute a routine behavior, one must then question the factors that govern these publication choices. At this level, however, the obstacle of research evaluation criteria, which date back to a time when OA did not exist, has been regularly highlighted. Encouraging individual incentives to publish therefore implies that other evaluation grids can quickly emerge. This is one of the objectives of the recent European Coalition S, which aims to generalize open access to all scientific communities:

We also understand that researchers may be driven to do so by a misdirected reward system which puts emphasis on the wrong indicators (e.g. journal impact factor). We therefore commit to fundamentally revise the incentive and reward system of science, using the San Francisco Declaration on Research Assessment as a starting point.

We also find this institutional will at the French level in the National Plan for Open Science[44] (2018, p. 4): "This evolution in the evaluation of researchers will aim to reduce the quantitative dimension in favor of a more qualitative evaluation, in the spirit of the San Francisco Declaration on Research Assessment (DORA) and the Leiden Manifesto[45] for the Measurement of Research, and by relying in particular on open citations, in the continuity of the efforts of the Initiative for Open Citations (I4OC)".

The increasingly strong injunctions against researchers in France to publish in OA can be interpreted as a way of getting around this obstacle in the short term. Until today, each researcher was free to choose and to publish as he or she wished. However, in Europe and in France in particular, institutions have shown unprecedented support for OA, which has resulted in a policy aimed at encouraging and then gradually obliging researchers to publish in this format. Within their Coalition S[46], they demand that all publications resulting from research work funded by public bodies must immediately be deposited in OA. In France, the recent National Plan on Open Science (2018, p. 4) proposed by the Ministry of Education and Research takes up this principle: "Publications resulting from research funded through calls for projects on public funds will be obligatorily made available in open access, either by publication in journals or books natively in OA, or by deposit in a public open archive such as HAL". To date, these principles have not been widely applied. It is the strategy that is considered the most relevant to achieve a sufficient critical mass of scientific documents in OA[47].

---

44 Available at: https://cache.media.enseignementsup-recherche.gouv.fr/file/Actus/67/2/PLAN_ NATIONAL_SCIENCE_OPEN_978672.pdf.

45 www.leidenmanifesto.org/.

46 www.coalition-s.org/about/.

47 With regard to the strategy to be adopted by the University of Aix-Marseille, the working group came to this conclusion: "All the experiments carried out over the last few years by organizations of different types and sizes worldwide show that, if we leave it at this stage, the

The other condition concerns the rate of use (reading) and reuse of the scientific documents contained in these self-archiving platforms. Indeed, the beneficial effects are often highlighted by the State, the research institutions and the creators of the platforms: reinforcing the visibility and scientific impact of research on a global scale by improving the dissemination of knowledge and research results; fighting against the oligopoly of scientific publishing and the impact of open access outside the scientific community (40% of the users of the platform in PubMed Central OS are citizens)[48]. All these objectives, however laudable they may be, do not directly concern the researcher's activity in the strict sense of the term. We can concede that a majority of researchers support these issues. However, defending OA does not necessarily imply an evolution of its publication practices. In the same way as for encouraging the deposit of scientific documents, their consultation and reuse are strongly linked to the routine practices of researchers. But while it is possible to force researchers to deposit the product of their research in an open archive to a certain extent, it seems to us much more delicate to oblige them to use it.

This is why it seems important to us, at this level, to clearly identify the reasons why a researcher goes to these platforms, to study the uses he makes of them, to evaluate which levers would be likely to increase these rates of use. At first glance, we can suspect that the denser these archives are, the more attractive they will be. In addition, these archives contain different types

archive filling rate rarely exceeds 10–15%, which is largely insufficient. The results of these same experiments converge towards a solution, which appears to be the most effective way to achieve optimal dissemination of the institutions' research output: the deposit obligation (or 'mandate'). The University of Liège and the University of Minho in Europe, as well as some thirty universities and *grandes écoles*, including Harvard and MIT in the United States, have chosen this option and thus manage to list practically all the articles produced by their researchers. In France, research organizations such as INRIA and IFREMER, as well as universities such as Angers and, more recently, Bordeaux, have set up or announced mandates to deposit their research publications. By adopting a similar mandate, AMU would be one of the first major French universities to display an ambitious policy for the promotion of open access" (Bertin *et al.* 2014). Available on HAL: https://hal-amu.archives-ouvertes.fr/ hal-01226882/document.

48 This information comes from a working document filed on HAL: Bertin, D., Dacos, M., Delhaye, M., Hug, M., Masclet de Barbarin, M. *et al* (2014). Vers une archive ouverte pour Aix-Marseille Université. Une démarche en faveur de l'open access: conclusions du groupe de travail sur le référencement des articles scientifiques produits par AMU [Online]. Technical report, Aix-Marseille University. Available at: https://hal-amu.archives-ouvertes.fr/ hal-01226882.

of scientific documents, preprints and postprints. The interest of the researcher is therefore not the same when looking for working documents and articles already published.

In the case of OA journals, the question is posed from a different angle for the researcher. Indeed, until recently, there was no question of publishing in an open or closed journal. At least, this was the case in many disciplines in the humanities and social sciences. The recent science policy aimed at supporting the development of OA in France through the National Plan for Open Science has certainly contributed to raising awareness on this issue. However, this has not yet translated into a willingness to give priority to OA journals, as shown by the recent barometer for open science[49], which does not show a radical change in recent years: between 2013 and 2018, the rate of open access to publications rose from 45% to 48.5%. Moreover, this rate varies significantly across disciplines (38.5% in social sciences and 71% in mathematics), which most likely means that the relationship with OA is conditioned by disciplinary practices, some of which have an older origin.

At a level that does not depend on the users but rather on the choices made regarding the governance of the platform, the enrichment of the SKC depends on the choices in terms of interoperability and the intellectual technologies they make available to researchers. Hess and Ostrom had mentioned the importance of socio-technical factors in the operation of an open archive and, in particular, the choices made in terms of the protocols chosen to promote interoperability between the different informational CPRs. These issues are being explored by some specialist researchers and information science experts. At this level, several questions arise, which also refer to the question of the governance of these platforms. Firstly, the enrichment of open archives is conditioned by their ability to link up with each other and, correlatively, by the existence of intelligent technologies that allow the exploration of all scientific data. In the case of HAL, for example, it is specified[50] that the deposit of articles allows the researcher to have an increased visibility because of the various partnerships (transfer of the deposits to the central PubMed archive for articles relating to the biomedical field), harvests carried out by other archives (Open Aire, the European archive, for example). Specialized search engines, such as Isidore, also allow harvesting

49 https://ministeresuprecherche.github.io/bso/.

50 www.ccsd.cnrs.fr/2018/03/moissonnage-et-referencements-de-hal/.

open archives, such as HAL in humanities and social sciences (HSS). But these approaches are still far from being exhaustive.

In this respect, Jean-Max Noyer and Maryse Carmes, while underlining the major importance of metadata to increase the quality and power of possible inferences to extract knowledge, processing modes, for example, statistical, allowing the exploitation of heterogeneous data, criticize the fact that these same metadata are too often thought of according to the paradigm of access and not in the perspective of supporting new modes of scientific production. The challenge is not to standardize the data catalogs of the different platforms: "Nothing would be worse than to essentialize the writings and approaches, to develop closed ontologies, at a time when the needs of governance must aim at processuality and the heterogeneity of actors and practices, criteria and ends, must aim at opening up to the differential relationships between micro-worlds and worlds" (Noyer and Carmes 2013, p. 6, author's translation). According to them, the production and dissemination of knowledge within information commons operate according to a bottom-up logic contrary to the top-down logic specific to constituted powers (political, religious, institutional, etc.) which gives rise to new "centered" modes of governance: "What signifies 'reuse of data' is the revival of the work process, the struggle for new democratic forms, new forms of creativity" (Noyer and Carmes 2013, author's translation).

It should also be noted that in the field of intelligent technologies, the practice of what is known as text and data mining (TDM), recognized as a major issue for science and research, is a practice that still suffers from a high degree of legal insecurity due to the lack of legal status, which prevents its deployment. This legal insecurity exists at two levels: the incompatibility of these exploration techniques with copyright provisions and the unsuitability of the database producer's right to dynamic processing of knowledge[51]. However, the practice of TDM is recognized as a decisive factor in revealing new research subjects and new knowledge and thus in responding to economic, social and societal[52] issues.

---

51 White Paper "Une science ouverte dans une République Numérique", 2016, p. 39.

52 One of the defenders of open content mining is Peter Murray Rust, a chemist, who defines this notion as "the unrestricted right of subscribers to extract, process and republish content manually or by machine in whatever form (text, diagrams, images, data, audio video, etc.) without prior specific permissions and subject only to community norms of responsible behaviour in the electronic age". In Murray-Rust, P., Molloy, J.C., Cabell, D. (2014). Open Content

It paves the way to opportunities in terms of valorization of this new knowledge with stakes of innovation, growth and employment. The Law for a Digital Republic introduced this principle to allow the "excavation" or "exploration" of "texts or data included in or associated with scientific writings" as an exception to copyright. However, the implementing decree has not yet been issued.

### 2.3.3.3. *Obstacles and threats to the development of scientific knowledge commons*

We refer here to a dense and very stimulating article recently written by Richard Poynder[53], which encourages us to present some of his arguments and to comment on them from our perspective of questioning the sustainability of common scientific knowledge in the current digital editorial ecosystem. A journalist by profession, Poynder has followed the OA movement since its beginnings and is considered by representatives to be one of the world's leading experts on this movement.

There are many obstacles today that jeopardize the development of OA at the international level. Some can be considered as exogenous, in the sense that they do not directly concern the digital editorial ecosystem. In particular, Poynder evokes geopolitical obstacles that constitute as many forces opposed to the expansion of OA[54]. Firstly, he mentions the populist and protectionist ideology of certain powerful countries that hinders the development dynamics of international scientific cooperation. He also mentions the Internet, which, by becoming a space destined primarily for the trade of personal data and simultaneously for forms of mass surveillance and digital[55]

---

Mining [Online]. In *Issues in Open Research Data*, Moore, S.A. (ed.) Ubiquity Press, London, 11–30. Available at: http://dx.doi.org/10.5334/ban.b.

53 "Open access: Could defeat be snatched from the jaws of victory?", November 18, 2019. Available at: https://poynder.blogspot.com/2019/11/open-access-could-defeat-be-snatched.html.

54 "In October it was reported that UK intelligence agencies MI5 and GCHQ had 'warned universities to put national security before commercial interest as fears grow over state theft of research and intellectual property from campuses. Both China and Russia were named, and universities were told that the growing number of international collaborations requires particular care'". In "Open access: Could defeat be snatched from the jaws of victory?", November 18, 2019, p. 50. Available at: https://poynder.blogspot.com/2019/11/open-access-could-defeat-be-snatched.html.

55 "The Internet Society reports that in 2018 freedom on the global internet declined for the eighth straight year, with a group of countries moving toward what it calls 'digital authoritarism'. Elsewhere, Freedom House reports that about 47% of Internet users now live in countries

authoritarianism (almost half of Internet users live in countries that restrict the conditions of access to social media), is moving inexorably away from its primary objective, which was to disseminate information and knowledge through OA.

Richard Poynder also points out the threats identified at the very heart of today's digital editorial. He strongly criticizes the growing tendency of journals in OA to adopt the APC economic model (publisher pays), which risks introducing new tariff barriers that are even more dangerous for research dynamics, since a world dominated by OA in APC would deprive a considerable number of researchers from being able to publish (and not only in Southern countries, as is often mentioned).

In addition, he points out that among the OA journals adopting this model, a number fall into the category of so-called predatory journals, which adopt publishing practices that do not respect the criteria of the scientific community such as promoting articles that are financed by industries[56]. Poynder points to the naivety of some advocates of OA who did not necessarily anticipate the deleterious effects of this type of economic model.

The second target of his criticism concerns the European Plan S which, according to him, indirectly reinforces the commodification of the editorial ecosystem with the risk of seeing new forms of enclosures appear. He criticizes in particular the injunction of Plan S to publish scientific documents in OA under a very permissive license, CC BY[57]. If the desired effect is to prevent *ex-post* appropriation of CC BY articles by traditional publishers, two other induced effects can be very detrimental. On the one

---

where access to social media or messaging platforms has been temporarily or permanently blocked" (*idem*, p. 50).

56 "Predatory publishers also allow dishonest researchers to scam their institutions and funders in order to advance their careers and obtain funding on false pretences... In 2010 three medical doctors, a biostatistician, and a research librarian examined the funding source and access status of 216 extended reports published between 2007 and 2008 in the *Annals of the Rheumatic Diseases*, a journal published by the prestigious BMJ Group. Their conclusion: 'Author-paid open access publishing preferentially increases accessibility to studies funded by industry. This could favor dissemination of pro-industry results'" (*idem*, p. 41).

57 Recall that this was also a recommendation already present in the BOAI: "We recommend the CC-BY license, or any other equivalent license, as the optimal license for the publication, distribution, use and reuse of academic works". See www.budapestopenaccessinitiative.org/boai-10-translations/french.

hand, this makes it possible for innovation actors in Northern countries to appropriate the knowledge to the detriment of Southern countries, which are not in a position to systematically exploit the economic potential of their own scientific research[58]. On the other hand, while such an option prevents direct privatization by publishers, it does not force them to exploit knowledge indirectly. Indeed, it is quite possible for an international publisher to retrieve and centralize all scientific articles published in CC BY on a single database and create new editorial products whose economic added value lies in the marketing of additional services such as search engines or high-performance data mining tools. Such a strategy would also respond to the now increasing research costs for a researcher faced with OA content that is widely dispersed on the web. Richard Poynder's conclusion is that the OA movement as it was imagined by the pioneers, with reference to the BOAJ, has not sufficiently taken into account the question of the cohabitation of OA in an editorial ecosystem dominated by powerful market forces, or the fact that governments in favor of OA see it above all as a powerful lever for establishing new dynamics of innovation and growth: "The open access, open science and other open movements have given too little thought to the fact that they have perforce to operate in a neoliberal world, a world in which commercial enclosure is a natural instinct and invariable endpoint, regardless of whatever high-sounding claims researchers might make about laying the foundation for uniting humanity in a common intellectual conversation and quest for knowledge"[59] (2014, 72). In the background, we find that the challenge is a fundamental aspect of the articulation between the world of the commons and the market economy which, until now, has regulated a very important part of the production and distribution of scientific communication. At this stage, Poynder's insight is essential, for it clearly shows that cohabitation already seems compromised, at least in the current configuration. Does this imply a medium-term disappearance of these scientific knowledge commons? What is certain is

---

58 Poynder refers to the Mexico City Declaration, where strong criticism has been leveled against this very permissive licensing principle: "Developing nations see this as very much a North/South issue, not least because they fear it will allow large legacy publishers based in the North to capture and monetise research published in the Global South. For this reason, the signatories of the Declaración De México advised authors and publishers in the region to abjure CC BY in favor of the CC BY-NC-SA223 license" (*idem*, p. 67). However, Coalition S continues to insist on the use of a CC BY or CC O license: their argument is that this type of license is necessary to prevent publishers from acquiring exclusive rights to the papers they publish.

59 *Idem*, p. 72.

that their promoters must take into account these obstacles, which indeed constitute so many threats to their sustainability. The question of open licenses is at the heart of this debate, as is that of OA funding.

As we come to the end of this section, we wanted to try to identify the questions that need to be asked, thus opening up essential avenues of study to meticulously extend this complex questioning.

## 2.4. Cooperative platforms as social commons?

In France, the research program initiated by the economist Benjamin Coriat is one of the current major contributions in the continuation of Ostrom's work on knowledge commons in the digital ecosystem. The originality of this interdisciplinary program is to strategically bring together the foundations of Ostrom's approach to knowledge commons and those of the social and solidarity economy (SSE). In this perspective, it closely associates the representatives of this current of thought, both academics and activists, to the spaces of reflection opened up by its program. If at the outset the reflections focused on discussions around emblematic knowledge commons, such as free software or the Wikipedia encyclopedia, they have gradually shifted to cooperative digital (service) platforms.

### 2.4.1. *A rapprochement with the social and solidarity economy*

Since 2013, Benjamin Coriat has been coordinating an inter-disciplinary research program on commons, in the framework of an ANR Propice[60],

---

60 This ANR brings together three partners: the *Centre d'Économie de Paris Nord* (CEPN, University of Paris 13), the *Centre de recherche en droit des sciences et techniques* of the Université Paris 1 and the IRD of the Université d'Aix-Marseille. The objectives of the project are to propose a state of the art on the tensions between intellectual property (copyright and patent) and commons, to show how and in what way the new intellectual commons differ from traditional forms of property, to identify recent trends in the strategies deployed by the actors and to propose one or more typologies of commons and the institutional arrangements on which they are based, to highlight the economic models that can guarantee their sustainability, and finally to suggest ways to establish an institutional context more appropriate to the proper deployment of creative and innovative activities. See for more details: http://anr-propice.mshparisnord.fr/. This ANR resulted in the publication of 29 studies and working papers, available online, and was concluded with the organization of an international colloquium on the theme "Property and commons".

which was later extended by the EnCommuns project[61] until today. This research has led to the publication of two collective works. The first, entitled *Le retour des communs: la crise de l'idéologie propriétaire* (Coriat 2015), proposes an extension of Ostrom's theory to the field of information, knowledge and culture. In the introduction, Benjamin Coriat specifies the purpose of this research program, which is to study "the sets of resources of a literary, artistic, scientific and technical nature whose production and/or access are shared between individuals and communities associated with the construction and governance of these domains" (Coriat 2015, p. 13, author's translation). In the same way as Ostrom and Hess, he uses the terminology of "knowledge commons" to designate sets of resources that are collectively governed in order to allow shared access: "Behind a common, there is a community, and for this prosperous community, there are rules" (Coriat 2015, p. 13, author's translation). In this perspective, collective action is also an essential dimension of a knowledge commons. We can already see here a distance with the conception of the commons of BCIS jurists, for whom the community does not occupy a high place in the definition of a cultural commons.

The second work, *Vers une République des biens communs?* follows a symposium organized in Cerisy in 2016[62] as part of the EnCommuns program. This time, it places the question of the commons in a more ambitious, socio-political framework:

> After the period of the great return of the commons – in the reality of the world as in that of research – we are entering a new period, a sort of new age of the commons... This new age, to put it in a word, is that of the rooting of the commons in society, of their extension to ever-expanding areas of social life and their perpetuation over time (Alix *et al.* 2016, p. 2, author's translation).

The aim is to identify the different forms of commons, to assess the economic conditions for their sustainability and their capacity to be a lever for

---

61 Begun in 2016, this project under the direction of Benjamin Coriat (CEPN, UP13) associates three partner teams: IRJS (Paris 1 Panthéon Sorbonne; leader: J. Rochfeld), CEPRISCA (Université de Picardie; leader: Aurore Chaigneau) and CREDEG (Université de Nice; leader: S. Vanuxem). For more details: http://encommuns.com/.

62 www.ccic-cerisy.asso.fr/bienscommunsTM18.html.

change in the transformation of society. Their approach is no longer limited to a specific field of commons, but is part of a global framework of understanding ranging from natural resources to information resources.

Finally, the originality of this research program lies in the clear willingness of its various contributors to bring Ostrom's theory of the commons closer to the social and solidarity economy. In the wake of this program, the Coop des Communs was associated with the program, an association that aims precisely at building concrete alliances between the social and solidarity economy and the commons:

> The association Coop des Communs brings together activists, militants, researchers, and entrepreneurs from the Social and Solidarity Economy (SSE), from the world of the commons and also from public stakeholders... We are willing to contribute to the construction of an ecosystem that puts together co-built Common and SSE, and also interested public stakeholders[63].

The members of this association reflect this recent orientation of this research program. They include researchers specializing in the communal sector, such as Benjamin Coriat, Fabienne Orsi and Laura Aufrère, researchers specializing in the social and solidarity economy, such as Hervé Defalvard, SSE actors such as the president of the credit cooperative, Jean-Louis Bancel, and finally intellectual activists from the communal sector, such as Lionel Maurel and Michel Bauwens.

Following the Cerisy conference, a special issue of the journal RECMA (*Revue internationale d'économie sociale*) in 2017 was devoted to this question, entitled "SSE and the Commons"[64]. For Philippe Eynaud and Adrien Laurent, "the theoretical approaches around the commons and the solidarity economy share the same observation. They have in fact the common point of underlining the aporias of a conceptual schema built solely around the opposition and/or complementarity between the market and the State" (Eynaud and Laurent 2017, p. 28, author's translation). For Hervé Defalvard, the SSE is the place of expression of what he calls the social commons:

---

63 https://coopdescommuns.org/.

64 Dossier "ESS et communs", *RECMA,* 2017/3, No. 345.

> The social commons are a marginal block of the neoliberal system, which is structured around the cooperative regulation of the economy by a social group.... They are formed around the core of the solidarity economy, based on public/communal partnerships in which local authorities play a major role, while involving small traditional local businesses (Eynaud and Laurent 2017, p. 49, author's translation).

So the challenge is to identify emerging forms of knowledge commons in the digital ecosystem that would participate in the values specific to the social and solidarity economy in order to ensure their protection and encourage their deployment. This new alliance offers the possibility of renewing the imagination of the associative community.

More recently, a new research program entitled Tapas[65] (There Are Platforms As Alternatives), managed by the *Centre d'économie de l'université Paris Nord* (CEPN) in conjunction with the Coop des Communs and supported by the *Direction de l'animation de la recherche, des études et des statistiques* (DARES), extends these questions from a more empirical point of view. Initiated in 2019, this project proposes to analyze the conditions of development and sustainability of certain service platforms designated as "cooperatives" because of their adherence to principles of social and environmental sustainability, such as Openfood France, oiseaux de passage, CoopCycle, Mobicoop, France Barter, Pwiic, Tudigo and Framasoft, which constitute their first study sample. An international symposium closed their survey in April 2020 with a presentation of their results.

Bringing the approach of the commons closer to the SSE approach is justified in that it opens up a new perspective for thinking, at the dawn of the 21st century, about new foundations for a renewed political economy where the union of the commons and the social and solidarity economy would offer a lever for the transformation of contemporary capitalism. By advocating such a creative alliance, they also distance themselves from the vision defended by the jurists of the Berkman Center, which is in line with the continuity of a liberal political economy as we have highlighted. While the

---

65 The project is described in detail on the CEPN laboratory site: https://cepn.univ-paris13.fr/tapas/.

commons are not intended to replace the capitalist economy, they nevertheless embody the possibility of its progressive transformation.

## 2.4.2. *Conditions for exploiting the social value created*

Let us now return to the question of the cultural commons in the digital ecosystem. We have already mentioned the fact that the operation modalities of the social value created by the participatory practices of voluntary contributors is a central issue conditioning the development of the cultural commons. Within the framework of the research program studied here, we would like to present the arguments of Michel Bauwens, founder of the P2P Foundation and author of several books on the post-capitalist economy and the society of the commons, who was associated fairly quickly with the movement initiated by Benjamin Coriat[66]. Indeed, the latter has positioned this question at the heart of his analysis of digital cooperative platforms. His approach is critical, as it aims to denounce the misdeeds of digital capitalism and, in particular, service sharing platforms that exploit, for their sole benefit, the usage value created by their contributors. Michel Bauwens provided an instructive cartography of how sharing platforms conceive of the valorization of voluntary contributors[67].

Some platforms are based on an "extractive" model of the emblematic value of a so-called "netarchic" capitalism, such as Facebook or Google. They are characterized by an economic exploitation of the social value resulting from the activities of voluntary contributors for their sole benefit. The latter can interact on these platforms, exchange and create information, but their actions remain controlled by a centralized infrastructure whose function is to extract value from these exchanges. Conversely, other platforms are based on a "generative" model of value, as they seek to create value for communities and commons, both socially and environmentally. Wikipedia and Linux fall into the latter category. Unlike Wikipedia or Linux, which invest capital in

---

66 He was one of the contributors to the book on the return of the commons, where he presents his participation in the Ecuadorian government's FLOK (Free/Libre Open Knowledge) project to establish a participatory process to design a transition strategy based on free and open knowledge. He also participated in the Cerisy symposium by presenting a paper on the theme: "Comment créer une véritable économie du commun?"

67 He addresses this issue in two of his books (Bauwens 2015; Bauwens and Kostakis 2017).

networks or platforms that promote social production, they do not involve value capture.

Even if there is an implicit contract between users and owners of netarchic platforms, with the former agreeing to have their data exploited economically in exchange for free use of the communication network, there is nevertheless a real form of exploitation, since there is no return of the exchange value to the contributing users: "Those who create value cannot live off it... A system that gives nothing back to those who produce value is unbalanced and creates a value crisis in society" (Bauwens 2015, p. 74, author's translation). A few pages later, he concludes as follows: "Centralized control of the peer-to-peer dynamic is comparable to an annuitant situation. When you make a profit on something you did not create, you are an annuitant. Even from a capitalist point of view, there is dysfunction" (Bauwens 2015: 75, author's translation).

Conversely, platforms that rely on a value-generating model are defined by their purpose, which is to serve societal goals first and foremost.

This does not imply that they cannot make a profit, but this must take second place to the social objective which remains their priority. This typology has the merit of making a clear distinction between the universe of platforms, which is 100% market driven, and that which participates in a commons logic. But it leaves open a subsidiary question: what are the safeguards that prevent such platforms from not being reclaimed and ultimately risking dissolution within netarchic capitalism?

Let us recall that the community of volunteer contributors co-produces commons open to reuse (as is the case in the open source domain). The social value created can therefore a priori be reappropriated by everyone, including a commercial enterprise. The use of Creative Commons licenses is particularly targeted in that they do not impose any criteria in terms of purpose of use for those who reuse the value created by volunteer contributors.

Michel Bauwens proposes as a partial solution the use of a new type of license, called "reciprocal", such as the peer production license that obliges any organization that wants to use an information commons for commercial purposes to contribute to its production or to pay for its use. The advantage of such a license is that it initiates an accumulation process in the production of this commons, thanks to which the individuals who contribute to it can make a living from it.

The value produced by the platform would remain within it, allowing for self-replication of the commons. Only companies or cooperatives owned by the workers would be able to use the production of the commons for commercial purposes. But all financial gains in the form of profits generated must be redistributed to the workers (contributors of the commons) if they have not contributed directly to this production.

On the other hand, for any company whose ownership and governance are private, the use of the commons, from a commercial perspective, is prohibited. For Bauwens, these licenses make it possible to build a bridge between the communal approach and the cooperative movement. He proposes a new expression as a symbol of this fruitful union, "open cooperativism": "Open cooperatives should use commons-based reciprocity licensing to protect against value capture by capitalist enterprises but also to create solidarity between the allied and generative coalitions" (Bauwens and Niaros 2019, p. 40).

As early as 2014, this idea of reciprocal licensing was proposed during an intellectual debate in the framework of the FLOK (Free/Libre Open Knowledge)[68] project of the Peer to Peer Foundation by a German activist, Dimitri Kleiner[69]. It was presented as an adjustment of the Creative Commons CC BY NC SA license. At that time, there was a debate about whether the commons economy should be locked into a totally non-market universe or whether it could introduce a hybrid economy.

But a certain number refused this last possibility at the risk of dilution in the market economy. But as others point out, if we restrict commons to a non-market universe, then this inevitably constrains their development and their capacity to respond to societal issues that may require the monetization of certain uses. The use of this type of license would thus make it possible to promote the development of cooperative enterprises, as Pierre Carl Langlais

68 The FLOK project was a project that aimed at the establishment of a "free and open knowledge society". The plan itself considers ways to move away from Ecuador's economic model, based on oil extraction, to one based on open and shared knowledge. For more details, see: https://framablog.org/2014/04/04/flok-society/.

69 Kleiner, D. (2007). Copyfarleft and copyjustright [Online]. July 18. Available at: www.metamute.org/editorial/articles/copyfarleft-and-copyjustright.

points out: "A virtuous circle is taking shape: the commons in question have the means to attract many volunteers and to share entire sectors of the economy: because of this increasing communication, it is in the interest of companies to switch to a cooperative system"[70].

However, in practice, very few projects have used this type of license, as their effective implementation is not without many problems. For Lionel Maurel[71], the criterion of organic reciprocity underlying this license is very restrictive since it is aimed at only a small number of entities: companies that belong to their employees and cooperatives. Moreover, in the latter case, the cooperative notion is also poorly defined. For example, in French law, there is no guarantee that cooperative production companies (*sociétés cooperatives de production*, SCOPs) fall within this perimeter. According to him, another possible solution could be to imagine basing the peer production license on the criterion of limited lucrativity[72] inherited from the SSE. This criterion reintroduces an "organic" logic in the appreciation of the use, allowing clear identification of the organizations that can claim it. The copyright license set up by the Coop des Communs could be a possible solution to extend it to the perimeter of digital platforms according to him: "The resources of the Coop des Communs are by default made available under a CC-BY-NC-ND license, but it has been decided that external entities will be exempt from prior authorization and royalties if they carry out a non-profit or limited profit-making activity" (Maurel 2018, author's translation). Compared to the Creative Commons license, it introduces a new dimension which is the purpose of the use and its context.

The issue of reciprocal licenses, whatever their format, is an essential legal element to link the world of the commons and that of the SSE. But if they allow us to settle the question of the boundaries between the commons and the commercial economy, they remain silent on the status of the

---

70 Langlais, P.C. (2014). Rendre aux communs le produit des communs: la quête d'une licence réciproque [Online]. Working paper. Available at: https://scoms.hypotheses.org/241.

71 Maurel, L. (2018). Coopyright: enfin une licence à réciprocité pour faire le lien entre communs et ESS? [Online]. March 2. Available at: https://scinfolex.com/2018/03/02/coopyright-a-licence-a-reciprocite-to-make-the-link-between-commons-and-so.

72 This is a criterion used by the tax authorities to grant tax deductions; the associations know if they are in the limited lucrativity in relation to the tax regime applied to them. Entities such as SCOPs, SCICs and ESUS companies are considered to be intrinsically in the sphere of limited lucrability because of their operating principles (this is what emerges notably from the definition of the SSE retained in the Hamon law).

commons platforms: should they not themselves be subject to the same rules as SSE platforms? The question seems to us essential in particular for those belonging to the hybrid economy field. To shed some light on this question, we will take a detour through the digital labor approach developed by sociologist Antonio Casilli, which will then lead us to evoke the central question of platform cooperativism in the case of cultural practices. Like Michel Bauwens, Antonio Casilli has come closer to the movement initiated by Benjamin Coriat. His theory on digital labor allows indeed an essential reflection on the governance of digital platforms oriented towards the sharing of cultural resources.

### 2.4.3. *Governance of cooperative platforms*

Antonio Casilli (2016) introduced the idea of digital labor to describe this service economy which was built on the exploitation of so-called "click" workers, who were not or only slightly paid (in relation to the work provided), also introducing new forms of social dumping. He was also invited to the Cerisy colloquium to discuss this issue. Several platform ecosystems characterize this digital labor: service platforms such as Uber; micro-work platforms such as Amazon Mechanical Turk or Foule Factory, which rely on a category of workers he calls "task proletarians" because of the very low remuneration (associated with a work contract) they receive; social platforms such as YouTube, Facebook and connected objects where, this time, contributors are not paid for the content they create and from which the platform will extract value.

These ecosystems are very different, but they are all based on the same concept of "work". They are activities that produce economic value (monetization, bids, purchasing, etc.), framed from a contractual point of view, subject to performance metrics and finally based on a parasubordination link (the employee is not in the formal sense but must nevertheless respond to external orders). For Casilli, this conception of work embodies platform capitalism:

> This way of condemning part of the global labor force to precariousness, while subjecting the other part to a leisure activity that produces value, comes from the same will that animates the capitalists of the platforms: to weaken labor in

order to better evacuate it both as a conceptual category and as a productive factor to be remunerated[73].

Do the creative commons platforms as described by the BCIS jurists also involve some form of exploitation of work? It should be remembered that platforms in the hybrid economy rely on the voluntary work of contributors, whose value created is then exploited by them. The answer to this question could be yes. However, according to them, what makes this type of sharing platform successful is that it initiates a form of informal social contract between the contributors and the owners of the platform based on a mutually agreed (or win–win) benefit. They recognize, however, that this very fragile informal contract is at risk of being broken if the platforms do not respect the norms and values of the community on which they rely to extract economic value. On the strength of this observation, they have not, however, provided effective solutions to mitigate this potential risk. Indeed, it can be assumed that since this balance is by nature unstable, it is the very sustainability of this hybrid economy that is at stake. What is at stake here is the issue of governance which has not been the subject of particular attention among the BCIS jurists. Conversely, it is a central issue raised repeatedly by participants in the EnCommuns research program. Applied to the specific case of social commons platforms, the question is whether they should be based on a specific form of governance that could avoid these potential risks of exploitation and unsustainability in the medium term.

The theory of platform cooperativism proposed by two American researchers, Theodor Scholz and Nathan Schneider (2016), presents in this respect elements of reflection that can undoubtedly advance this question. Highly critical of platform capitalism (such as Bauwens), they propose an alternative, more "ethical" model and affirm their desire to revive the foundations of the social and solidarity economy, materialized by an alliance between the heritage of cooperatives and 21st century technologies.

Their approach focuses on service platforms, as well as extends to a few examples in the field of culture (music, photography). This is why we think it is also interesting to mention it here. Indeed, we may wonder whether this

---

73 Casilli, A. (2016). Digital labor: conflits et communs à l'heure des plateformes numériques [Online]. Video. Available at: www.colloque-tv.com/colloques/vers-une-republique-des-biens-communs/digital-labor-conflits-et-communs-%C3%A0-lheure-des-plateformes-num%C3%A9riques.

platform cooperativism might not correspond to one of the possible manifestations of cultural commons in the digital ecosystem. BCIS jurists have insisted a great deal on the use of free Creative Commons type licenses to define the perimeter of a cultural commons, but, as Benjamin Coriat reminds us, a commons, whether natural or intangible, defines "sets of resources that are collectively governed, by means of a governance structure that ensures a distribution of rights among the partners participating in the commons" (Coriat 2015, p. 38, author's translation). The question of governance is just as important as the question of bundles of rights of use of the commons.

For Nathan Schneider and Theodor Scholz, a cooperative platform, sometimes also called "platform commons", is materialized by a governance based on shared and democratic ownership. The data flows that often form the core of the value must be transparent, especially for contributors who need to understand the parameters and models that govern their working environment. Unlike current black box systems, such platforms make the storage locations of user and worker data, the terms and conditions of sale, and the beneficiaries transparent. Such a platform must also be committed to respecting the prevailing legal rules for work. The remuneration of contributors is at the heart of the issues at stake, since the objective is that through a democratic mode of governance, the excesses of capitalist platforms in terms of remuneration should be avoided: "The history of cooperatives in the United States has taught us that they are indeed able to offer a more stable income and a dignified workplace" (Scholz and Schneider 2016, p. 25).

This approach was inspired by the social movement of English cooperativism in the mid-19th century to which Nathan Schneider and Theodor Scholz refer explicitly, materialized by the "Rochdale Principles": open and voluntary membership to all regardless of their economic background; democratic control of the cooperative materialized by the right of each member to vote on the principle of one man one vote; social responsibility of the community through which members work for the improvement of society[74].

---

74 "One can trace the modern cooperative movement to the Rochdale Principles of 1844, in England, though it had precursors among ancient tribes, monasteries, and guilds around the world. The rudiments of this stuff could be basic common sense: shared ownership and governance among people who depend on an enterprise, shared profits, and coordination among enterprises rather than competition" (Schneider and Scholz 2016, p. 15).

At this stage of their reflection, they have not yet delivered a more assertive and precise theoretical grounding on platform cooperativism. One of their merits, however, is that they have presented a few case studies which, for them, are emblematic of this new trend that is taking shape in the digital sharing economy. This mapping deserves to be highlighted, because among the examples proposed, some platforms are part of the digital cultural economy.

Platform cooperativism mainly concerns service markets. This is why the question of fair remuneration for those who offer their services is central. Such platforms already exist in various fields. But they are still in limited number, their growth being strongly conditioned by the parallel deployment of a financial ecosystem supporting these start-ups with cooperative status in their launch and growth phase. In their 2016 book, Schneider and Scholz refer to a dozen projects worldwide that claim to belong to a form of platform cooperativism. They cite, for the most well-known, Fairmondo[75] in Germany, which aims to be a global online marketplace alternative to Amazon and eBay, Fairbnb[76], a vacation rental platform, the Openfood platform in the field of short food circuits, or Loconomics[77], which allows freelance workers to find clients in the San Francisco Bay[78] Area.

Among these platforms, two are of particular interest for our study because they directly concern the field of culture. These are the music streaming platform Resonate, launched in beta version in 2016, and the platform Stocksy United photos, created in 2012 in Canada. We can also add to this sample 1DLab, another French fair trade streaming platform created in 2013[79].

The two music streaming platforms have in common the desire to offer a more advantageous alternative to the methods of distributing the value of the

---

75 www.fairmondo.de/.

76 https://fairbnb.coop/.

77 https://loconomics.com/.

78 In France, too, there are such cooperative platforms. The Coop des Communs has prepared a report on this subject which is available online: https://coopdescommuns.org/fr/la-communaute-plateformes-en-communs-fait-son-point-detape-article/.

79 This platform is part of the structures participating in the Plateformes en communs project by the Coop des Communs. See: https://coopdescommuns.org/fr/la-communaute-plateformes-en-communs-fait-son-point-detape-article/. The platform's website is available at: http://1d-lab.eu/.

main players in the music distribution market (Spotify, iTunes, Deezer, etc.). They therefore address professionals, artists and producers (independent labels) directly, promising them fairer remuneration, which is more advantageous for them compared to traditional market platforms.

In terms of the economic model, they propose original solutions aimed at reintroducing the idea of a willingness to pay. For Resonate, their *Stream2own* model is based on the "listen to own" principle. The listener is obliged to create an account that is immediately credited with free listening time, but which he/she must then top up with a minimum of five euros. In the discovery phase, each first listening costs only a few cents, that is, a hundred songs listened to for 2 to 3 euros. Then, if the listener listens to the same stream more than five times, he/she enters the "fan" phase and the prices start to be more substantial. After the ninth listening, the song is available for free forever (no downloading possible). The monthly cost for the listener is about the same as the average subscription amount available on standard platforms. On the other hand, the remuneration system is more advantageous for the artist, who receives for nine streams what he/she should receive for 100 streams on a standard[80] platform. Finally, Resonate uses blockchain technology to build a secure and transparent database to manage the streams.

1DLab has signed agreements with a number of digital distributors (Believe Digital, Idol, CD1D, etc.) which have enabled it to have a fund of approximately 1 million titles. It sells subscriptions to intermediary structures such as libraries, which then offer access to their offerings to their members. This subscription, the "territorial creative contribution", is collected from the various partners and redistributed according to a mechanism that promotes cultural diversity. Revenues are distributed among the rights holders as follows: a fixed portion (15%), a variable portion (40%) (linked to the number of listenings) and a solidarity savings fund that can be used to supply producers and finance collective projects. The remainder is allocated to internal operations (20%) and R&D (15%). The status of the platform is an SCIC (*société coopérative d'intérêt collectif*, a cooperative society of collective interest). It brings together all the actors of the creative ecosystem. Governance is shared: 35 members divided into five colleges (creators, distributors, employees, groups and federations and support members). There does not seem to be any redistribution of profits.

---

80 https://medium.com/resonatecoop/streaming-sucks-ac585808b8a6.

For its part, Resonate also presents itself as a cooperative (legal type not specified) multi-stakeholder, where musicians, fans and staff share benefits and roles within the governance structure. For voting procedures, the structure is composed of one member and one vote, while profits are distributed according to the value generated by the participants. The different values are the amount of flow between musicians, the expenses of the listeners and the time commitments of staff and volunteers. To date, Resonate has approximately 4,000 titles.

From a copyright point of view, these music platform projects do not offer their music in the form of a Creative Commons type free license allowing the sharing of works with a minimum right of copying in the possible uses. In fact, these platforms do not directly address amateurs but rather professionals via their intermediaries, the labels. They open up a new and unprecedented perspective in the traditional musical world, relying on the originality of their economic model and their cooperative grounding to attract a sufficient community to allow them to survive in a dominated world.

Can these cooperative platforms in the field of culture be considered as new forms of cultural commons? This is a question that does not call for a binary answer because, as we have pointed out, it is difficult to offer a single reading key for the definition of a cultural commons in the digital ecosystem. We would say that the answer depends on the conceptual framework in which we place ourselves.

If we place ourselves in the conceptual framework of the Berkman Center's jurists, the answer is negative: the music is protected by copyright and the artists will give up their rights exclusively to the platform to be able to distribute them according to rules set by the *Stream2own*[81] model. The same is true for 1Dlab. This type of platform is not in the field of free culture

---

81 "All content provided by our users for the service is provided on the basis of a non-exclusive license.... By uploading music to Resonate, the member grants Resonate Cooperative a revocable license to distribute content on the member's behalf via streaming and digital download in accordance with the #stream2own model.... A musician member can revoke this license at their discretion by simply disabling a track from being streamed. Disabled tracks will remain in our system until manually deleted by the content owner. However, the track will remain available for digital download by all members which own the track via #stream2own for 12 months from the start of the notice period" (source: https://resonate.is/terms-conditions/).

with the sharing of cultural works between amateurs, but is primarily aimed at traditional players in the value chain of cultural industries.

If we place ourselves in the Ostromian framework, we might be tempted to answer in the affirmative, because each of these platforms is presented as an entirety of resources (in this case music) which are collectively governed, with a clearly identified governance structure associating not only the founders of these platforms, but also users (listeners) and "content providers" by assigning them specific rights in terms of use, appropriation and collective management. The commoners, identified as the artists (as well as the listeners in the case of Resonate), participate in the joint management of all musical resources. We are fairly close to the forms of land commons, which are communities that are open to their members, but closed to the outside world. These forms of cooperatives resemble what Michel Bauwens calls closed commons:

> Classical cooperative models still function as "private property" in relation to external commons and can at best create "closed commons". It is therefore vital to develop new forms of cooperation in which the creation of open commons is constitutive of their objectives and activities, both with regard to the immaterial commons of knowledge and the mutualization of their physical infrastructures (Bauwens and Niaro 2019, p. 40).

At the beginning of this section, we introduced the digital labor theory, seeking to assess the extent to which it also applies to the cultural sphere in the digital ecosystem. This question deserves to be asked and invested in more closely in the case of hybrid economy platforms that can drift, like any other digital platform, towards an excessive exploitation of voluntary contributions to the creation of value. This has led us to introduce platform cooperativism, presented by its architects as one of the solutions to mitigate the deleterious effects of capitalist platforms. Some rare cultural platforms have started this shift by inscribing their approach in the field of the social and solidarity economy. Their belonging to the world of cultural commons is not an easy question to solve. Beyond that, it reveals how difficult it is to have a single reading key to identify what is called a cultural commons in the digital ecosystem. Should priority be given to the criterion of individual choice for sharing the resource (in the spirit of free culture), or should collective action and governance of the common cultural resource pool be considered first and foremost? We might be tempted to say that among the platforms offering cultural content, two categories of cultural commons should be considered: creative cultural commons in the spirit of free culture and cooperative cultural commons.

### 2.4.4. *Commoners' remuneration: a right to contribute*

Before closing this chapter, a fitting final space for reflection seems to be the extensions of the Ostromian approach in the field of knowledge and culture by the proponents of the research program initiated by Coriat. On many occasions, the need to raise the question of the remuneration of commoners as one of the ways of perpetuating the commons has been stressed. This question is of concern to all research programs on the commons today, as we have seen in the work of De Filippi in the framework of his research at the Berkman Center. We will see here that the solutions envisaged are in some way part of the continuity of the search for proximity with the field of the social and solidarity economy.

Unlike the debates on platform cooperativism, which are not directly addressed to the field of cultural commons platforms, the reflections presented here are openly part of the research program on commons initiated by Coriat in particular. As an example, we can cite the seminar of the EnCommuns program held at the MSH on May 16, 2019, which focused on the theme "How do we pay commoners?"[82] One of the current challenges for the sustainability of commoners is based on the economic situation of those who, through their cooperative or collaborative action, create economic and social value. However, as their contributions are voluntary, there is a medium-term risk of the disappearance of these creative dynamics of collective commons in the field of knowledge and therefore particularly in the cultural field. The development of a commons economy cannot be based solely on voluntary contributions.

While at the present time, to my knowledge, there is no clear and assertive answer to this question, a real project has been launched, starting with the work carried out directly by Benjamin Coriat within the framework of the think tank of the group Appalled Economists around the idea of

---

82 http://encommuns.com/.

"common labor rights"[83], and then in a more distant way, by intellectuals, some of whom are close to the world of the commons, such as Michel Bauwens and his idea of the sovereignty of value. Contrary to the Sabir system put forward by De Filippi, which is part of an individualistic logic of individual reward, the underlying logic is different here, the objective being, first and foremost, the survival and durability of the common before individual satisfaction.

More than a reward, it is a right to the contribution that everyone must obtain. This right does not have to be individualized. It must be the same for everyone on an egalitarian basis. Thus, Coriat introduces the idea of common labor rights inspired by the thought of the jurist Alain Supiot[84], who broadens the conception of work by distinguishing it from the notion of employment to which it is usually attached. Thus, not only paid activities, but also non-market activities considered socially useful (lifelong training of individuals, raising children, caring for the elderly or sick, volunteer work through associative or citizen commitments) give rise to "social drawing rights". For Coriat, the idea of "common labor rights" should include, alongside these non-market activities, the contribution to common areas (be they urban, territorial, land, informational or cultural).

This idea of the right to contribute is in line with other proposals put forward proposing mutualized funding to remunerate voluntary contributors and as the basis for a contribution economy. For example, Michel Bauwens and Visalis Kostakis envisage a possible evolution of capitalism where markets would be based on a generative conception of value, "under civic domination" (Bauwens and Kostakis 2017, p. 83), in the service of the commons. One of the essential foundations of such a society of the commons is the establishment of a basic income that would develop in parallel with existing welfare models and that would allow the income from activities in the commons (locally or globally) to be supplemented with the use of the reciprocal licenses already mentioned.

We can also mention Bernard Stiegler's proposal of contributory income, even if the starting point of his analysis does not concern digital commons

---

83 Les économistes atterrés (2017). *Changer d'avenir. Réinventer le travail et le modèle économique*. Les liens qui libèrent, Paris. Edited by Mireille Bruyère, Benjamin Coriat, Nathalie Coutinet, Jean-Marie Harribey.
84 Supiot, A. (ed.) (2016). *Au-delà de l'emploi*. Flammarion, Paris.

directly and voluntary contributions, but the observation of a widespread automation of company creation, leading to the disappearance of the paid worker. An opportunity is emerging for "deproletarianized" workers to devote their time to contributory activities based on collective intelligence. But for this to happen, a new system of redistribution, contributory income and a new form of work organization must be introduced.

Right to contribution, basic income and contributory income are all proposals that aim at a profound transformation of current capitalism by allowing the extension of a world of commons (natural, urban, digital) and a reversal of social values by giving priority to solidarity and equality. However, not all these proposals for financing the voluntary contribution are equivalent and, for a certain number, still in a state of gestation. However, even if it is still difficult to have a precise idea of the political economy behind each of these proposals, it is certain that they have a common enemy, the attention economy as the new playground of a creeping digital capitalism. There is no question here of cohabiting commons with capitalism as it is. It is a question of creating the conditions for the development of the commons to destabilize its foundations and transform it from within. It is not the vocation of the commons to replace the market, but to modify market agencies.

PART 2

# The Commons in the
# Digital Book Ecosystem

# Introduction to Part 2

In spite of a difficult recognition of the cultural commons in the legislative institutional space, creative dynamics drawing the outlines of a cultural economy of the commons have gradually emerged in search of sustainable paths of development. The digital ecosystem is made up of a complex interlacing of information technologies that bring innovations, new information and communication uses and new markets seeking to exploit this new area of economic development. These multiple forms of interlacing will gradually change the conditions for the production, circulation and sharing of knowledge and culture.

Creative practices emblematic of free culture, likely to give rise to these cultural commons in the digital ecosystem, are fueling the transformation of existing commercial arrangements. They challenge the market logic of the cultural industries by questioning some of its fundamental data. The emergence of the amateur, already identified as being at the heart of this new ecosystem, is blurring the boundaries between producer and consumer at the basis of the traditional logic of markets. Moreover, when the amateur chooses to share his or her cultural creation (in the field of sound, image or text) in the form of free access, or chooses to contribute voluntarily to a collective social production, he or she inevitably disrupts the environment of the actors involved in the cultural markets, but without necessarily having the intention to do so. He or she also contributes indirectly to the emergence of new forms of intermediation with platforms dedicated to the expression of these individual or collective practices. These amateur cultural practices have the potential to renew creative processes and innovation, as highlighted by the different theoretical approaches on cultural commons developed in Part 1 of this book.

However, the paths taken by the actors promoting this free culture are multiple, which makes the outlines of the economy of the cultural commons inevitably complex to grasp at first glance. The nature and extent of the transformations of commercial arrangements in the cultural sphere remain to this day a work in progress on which we will shed light. The choice was made to confine this exploratory study to the field of books alone, which is approached within its two complementary universes: libraries and publishing.

For each of the worlds studied, we propose first to review some of the transformations underway in the digital book ecosystem, which have created the conditions favorable to the institution of new forms of cultural commons. We will then turn our attention to the conditions for the deployment of this cultural commons economy.

In the world of libraries, the institutional, legal and economic conditions favorable to the emergence of heritage commons will be examined in the light of some emblematic illustrations of digitization projects initiated in France.

In the field of publishing, we propose to identify the potential for deploying a commons economy of the written word through several case studies of platforms hosting cultural content from free amateur practices.

# 3

# Digital Libraries as Heritage Commons

Hess and Ostrom hypothesized that digital archives and libraries, particularly in the scientific field, were eligible for knowledge commons status. We then tried to show here how their approach needed to be compared to open access initiatives and, more broadly, to the transformations of the scientific publishing environment in order to better highlight their potential contribution. We have now chosen to study this question in another context related to the challenges of digitizing cultural heritage for public libraries (in France). To our knowledge, this is an issue that has not been the subject of any particular investigation. Under the term "cultural heritage", we refer here to all books and journals available in libraries that are in the public domain.

The digital ecosystem offers an unparalleled opportunity to make public domain works available to everyone, not only through shared access, but also with the possibility of reuse by third parties in the form of digital libraries. Physical libraries, each of which holds a part of this heritage and a large part of which often remains inaccessible to the public, are at the heart of this new challenge of building up commons potential that we choose to call "heritage".

Deciding on the eligibility of this digital cultural heritage for commons status requires, beforehand, that a reading grid be mobilized on the basis of the contributions resulting from the different approaches to cultural commons presented above. Ostrom's approach seems to us to be a relevant starting point, as digital cultural heritage can be similar to a common pool resource (CPR). On the contrary, unlike CPRs in the form of scientific digital archives, this CPR is the result of a production process stemming from the digitization strategies of the different entities producing the resource units constituting this CPR, represented here by the digitized content. Thus, the question of financing the constitution of such digital archives will appear as one of the fundamental issues related to the development of these heritage CPRs.

The core of our objective will be, from a microsituational perspective, to locate and identify the different institutional rules (legal, economic and managerial) enacted by the actors involved in the production process of digitized cultural heritage. As we will see, this production is always based on hybrid production logics that are neither exclusively market-driven nor publicly funded. Before doing so, we would like to mention some external factors that constitute a favorable ground for the emergence of these forms of heritage commons.

## 3.1. A favorable context

### 3.1.1. *A new documentary order*

The unprecedented possibility of creating a universal library that would bring together, in digital format, all published books, all printed and manuscript texts, following the example of the mythical figure of the Library of Alexandria, very quickly became a tangible reality. As the book historian Roger Chartier points out: "The communication of texts remotely, which cancels the hitherto distinction between the place of the text and the place of the reader, makes this ancient dream possible and accessible."[1]

Pioneering initiatives have helped to make this myth tangible, such as Project Gutenberg, launched by computer scientist Michael Hart in 1971, which is considered to have given rise to the first digital book. It was in fact the transcription on his computer of the Declaration of Independence of the United States, deposited on a server on the Arpanet network for free download. As Hervé Le Crosnier points out[2], this desire to make works available to all was inspired by the idea of a universal library launched by the "founding fathers" of computer science (Vennevar Bush, Joseph Licklider and Ted Nelson). While, at the beginning, this was a solitary project, it gradually became a collective project mobilizing millions of people over time. Its instigator, Michael Hart, had in fact quite quickly launched an appeal to all

---

1 Chartier, R. (1994). Du codex à l'écran : les trajectoires de l'écrit [Online]. *Solaris*. Available at: http://gabriel.gallezot.free.fr/Solaris/d01/1chartier.html.

2 Le Crosnier, H. (2011). Le projet Gutenberg est orphelin [Online]. *Le Monde diplomatique*. Available at: https://blog.mondediplo.net/2011-09-11-Le-projet-Gutenberg-est-orphelin.

volunteers wishing to make printed books, most of them in the public domain, accessible in the form of electronic text. By 2011, the library had collected 37,000 books in 60 languages. The originality of this project is twofold. First, it reveals that physical libraries are not the only legitimate actors in the construction of digital libraries. Second, Project Gutenberg makes content available to everyone for free and makes the digitized works freely reusable (even commercially)[3]. It thus introduces the possibility of a cultural heritage as a commons.

The fact that new players, from outside the world of libraries, have positioned themselves primarily on such projects can be explained by the fact that the digital ecosystem is, in itself, a new self-managed documentary order based on new rules that are not identical to the traditional classification rules of the librarians' documentary order. As Michel Salaün points out, the originality of the new documentary order consubstantial to the Web is to introduce direct links between the elements contained in documents, without going through a classification or index specific to the world of librarians: "In this conception, the document is no longer just an object, but also a node or a network head that makes it possible to move from one document to another" (Salaün 2012, p. 72, author's translation). A document (text, image or sound) can be described in a language that allows communication between the different "places" of publication thanks to what is called metadata, essential data on which search engine algorithms are based. Nowadays, since publication is done directly on the Web, the documents are there; the challenge is to make them accessible to the reader. The indexing of the digital document must therefore be thought of simultaneously with its construction so that it can be found. With the emergence of the Data Web, or Semantic Web[4], marking the second stage of development of the Web, the challenge is "to provide a language that expresses both data and rules for reasoning about data and so that the rules of any knowledge representation system can be exported to the Web" (Salaün 2012, p. 73, author's translation).

---

3 For more details, see Lebert, M. (2009). Le projet Gutenberg (1971–2009) [Online]. Available at: www.gutenberg.org/files/31634/31634-pdf.pdf?session_id=4a57e1c6bbbdc0ef6af8 b5d92fbbf8c7321cf724.

4 The objective of the Semantic Web is to propose standards that allow everyone to structure their productions, especially textual ones, using taxonomies or ontologies and formalisms such as RDF (Resource Description Framework) or OWL (Ontology Web Language). These technologies aim to standardize knowledge modeling.

This transformation of the documentary order is an essential factor to be taken into account in any digital library project, which cannot avoid opening up a debate on the modalities of digital archiving, data format and interoperability, as Milad Doueihi points out: "Interoperability and compatibility are therefore the basic characteristics of any conceivable framework for digital archiving, and this requirement becomes a strong argument in favor of open formats" (Doueihi 2008, p. 212, author's translation).

However, we should not believe that everything is so simple. There are many modalities of digital archiving; the digitization of the sign is a process that lends itself to the creation of plural intellectual technologies. This is the idea defended by Brigitte Juanals and Jean-Max Noyer under the expression of digital plasticity. They insist on the fact that the conditions of digitization of the sign can produce open and dynamic rhizomic informational and communicational networks as well as tree-centered networks. "In this context, the development of new forms of memory and distributed intelligences attached to the process of digitization of the sign constitutes a major 'strategic environment'" (Juanals and Noyer 2010, p. 29, author's translation). The conditions of production, circulation and consumption of knowledge are therefore strongly conditioned by the nature of the intellectual technologies produced by these processes involving the digitization of the sign (text, image, sound). The major challenge of an economic and political nature is to promote those that will enable the activation of collective intelligence in the sense given by Pierre Levy. Let us recall that for the latter, the pooling of imaginations and knowledge must "promote the constitution of intelligent collectives where the social and cognitive potentialities of each person can develop and mutually amplify" (Levy 1994, p. 30, author's translation).

The properties of information infrastructure appear as fundamental data of any project constituting a digital library, as well as of the possibilities which are offered to it to be able to be enriched and to give place to multiple creative activities. In this sense, we can say that they also condition the deployment of heritage commons.

### 3.1.2. *Cultural public data as a public good*

The interest shown by public authorities in the challenges of digitizing digital heritage was seen very early on, through the support given to the various initiatives undertaken by libraries, and in particular by the Bibliothèque nationale de France (BnF), whose first digitization program dates back to the end of the 1990s. However, it was from 2012 onwards that we can detect a strong desire on the part of the government to promote access to digital cultural content that has entered the public domain. It had already shown itself to be in favor of opening up public data, through its open data policy[5] led by the Etalab mission. This policy aimed to go beyond a policy of access to administrative documents to encourage public and private dynamics for reusing these same data. This new public policy was later extended to the cultural domain.

This policy was strongly influenced by the Domange Report (named after its rapporteur) commissioned by Minister Aurore Filippetti in 2013[6]. This report helped to reveal the strategic importance of opening up cultural data, defined in three ways: 1) statistical and economic data held by cultural institutions; 2) cultural metadata; and 3) digital resources in the form of image files or digital copies of works that have entered the public domain.

The opening up of cultural data is considered an essential lever to the emergence of creation and innovation ecosystems. The report states that cultural institutions must now adapt their mission to this new digital environment "by learning to introduce the concept of 'hackability' into their organization, in order to successfully integrate innovation from outside and capture external creativity"[7]. This opening also represents a new opportunity to promote cultural democratization and the transmission of knowledge by restoring direct links with the user. In this perspective, they are then related to new forms of common goods: "These data that participate in the education of citizens and the youngest in society, that promote cultural democratization and the transmission of knowledge are true common goods and are part of the cultural and historical heritage of every citizen"[8]. The choice of this terminology is timely, because it opens up and legitimizes a reflection on the institutional governance of these open cultural data and the emergence of new

5 Public data are data appearing in documents produced or received by administrations within the framework of a public service mission, regardless of the medium. Data can be stored and classified in different forms: paper, digital, alphabetical, images, sounds, etc.

6 Domange, C. (2013). Ouverture et partage des données publiques du secteur culturel. Pour une (r)évolution numérique dans le secteur culturel. Report, Département des programmes numériques du Secrétariat Général, December.

7 *Ibid*, p. 2.

8 *Ibid*.

forms of cultural commons. The Domange report sheds some light on this question by specifying the two levels of use that can be made of public cultural data: simple access (1) or reuse (2):

> (1) "The free and open reuse of basic or raw cultural data that has not given rise to any work of enrichment, contextualization, editorial added value or structuring that indicates the producer's intention to give them a cultural or cognitive meaning or appropriation."

> (2) "The multiplication of free access to public digital data in the cultural sector through the development of widgets that allow querying a programming interface (or API) that allows *queries to be made within the data, without the need to download the entire data.*"[9]

This desire to support the opening up and reuse of digital cultural resources has resulted in a clarification of the legal framework concerning the reuse of public data in the cultural sector and, in particular, a strong incentive to use open licenses. As such, the Domange report states that cultural institutions should resist the urge to create tailor-made license agreements and promote the use of standardized licenses that allow for the determination of the reciprocal rights and obligations of the parties.

As book cultural heritage is in the public domain, the intellectual property code does not recognize a positive definition of it. However, there are now open legal licenses that allow it to be formally identified and therefore protected against any attempt to enclose it. These are, on the one hand, the Public Domain Mark (PDM) license created by the Creative Commons association, and, on the other hand, the Etalab license, a free French license created by the Etalab mission[10] as part of its mission to support the release of public data. The PDM license certifies in a distinctive way that a work

---

9 Domange, C. (2013). Pour une stratégie numérique de diffusion et de réutilisation des données publiques numériques du secteur culturel. *Guide Data Culture,* 2013(1), 40.

10 As part of the government's policy to open up public data, it has been proposed to use the Etalab Open License, which is designed to encourage the reuse of public data made available free of charge. This license is compatible with other licenses of the same type at the international level and in particular with the CC BY 2.0 license. Etalab is the name given to the government mission that, since the end of 2017, has been coordinating the policy of opening up and sharing public data (open data).

belongs to the public domain. It makes a distinction between the "creator" and the "curator", that is, between the author of the work that has fallen into the public domain and the institution holding the original that is carrying out the digitization. It is the latter that "marks", as its name indicates, the digitized work to certify that it belongs to the public domain. The Etalab Open License is in line with the provisions of the 1978 law[11]. On an international scale, it is compatible with the standards of open data licenses developed abroad, in particular those of the British government (Open Government Licence) as well as other international standards (ODC BY, CC BY 2.0). The main elements of this license[12] are presented in Box 3.1.

---

The "licensor" grants the "reuser" a non-exclusive and free right to freely "reuse" the "information" subject to this license, for commercial or non-commercial purposes, worldwide and for an unlimited period of time, under the conditions expressed below.

The "reuser" is free to reuse the "information":

– to reproduce it and copy it;

– to adapt, modify, extract and transform it, to create "derived information", products or services;

– to communicate, disseminate, redistribute, publish and transmit it;

– to exploit it commercially, for example, by combining it with other information, or by including it in its own product or application.

Subject to the requirement to acknowledge authorship of the "information": its source (at least the name of the "licensor") and the date on which the reused "information" was last updated.

The "reuser" can in particular fulfill this condition by referring, by a hypertext link, to the source of the "information" and ensuring an effective mention of his authorship.

---

**Box 3.1.** *Content of the Etalab Open License*

---

11 This law defines the legal framework for the reuse of public data in the cultural sector. It specifies the scope of so-called public data and the conditions for its reuse.

12 www.etalab.gouv.fr/wp-content/uploads/2017/04/ETALAB-Licence-Ouverte-v2.0.pdf.

In reality, however, this promotion of a voluntarist policy in favor of the reuse of public cultural data[13] has come up against the recurrent use of royalties by a number of cultural institutions. Several reasons have been cited. Among them, some institutions mention the fact that the digitization process has a significant cost for them, and they have no other solution than to pass this cost on to the user. The Valter law of December 28, 2015, corresponding to the transposition of the European directive on the reuse of public sector information and in line with the law for a Digital Republic of October 7, 2016, proposes to clarify this stumbling block. This law sets a new framework for the use of cultural data, with the general rule of encouraging the reuse of public information by broadening the scope of the right to reuse. Now, cultural institutions and services (i.e. archives) and educational and research institutions are subject to the general law on reuse (whereas they previously belonged to the derogatory perimeter defined in the former article 11 of the 1978 law). It also establishes the principle of free access as a standard. However, it has not, for all that, prohibited the levying of a fee. This possibility now falls within the regime of exceptions. It is authorized in two cases[14], one of which concerns the reuse of information resulting from the digitization of library collections, including university libraries, museums and archives, and, where applicable, associated information when the latter are jointly marketed.

It can be seen here that in this law, the objective of opening up public data is not considered to be totally incompatible with the implementation of a financing system in the form of royalties. But this can only be done under certain very strict conditions. The digitization projects of library heritage are primarily concerned by this exceptional scope. In this case, it is specified that the total amount of this fee may not exceed, over a given period, the total amount of costs related to the collection, production, availability to the public or dissemination of their public information. It is stipulated that an agreement with a private company may be concluded, giving rise to a temporary privatization of the resource for a period that may not exceed 15 years. However, a copy of the digitized resources and associated data is provided free of charge, in an open and easily reusable standard, to administrations. Finally,

---

13 The reuse of public information is a use by a third party for purposes other than those of the public service mission for which the documents were produced or received.

14 The other case concerns public administrations that are required to cover a substantial part of the costs related to the fulfillment of their public service mission with their own revenues (e.g. IGN, Météo France).

with regard to the editorial documentarization process, this law specifies that "when this information is made available in electronic form, it shall be made available, if possible, in an open and easily reusable standard, i.e. machine-readable" (Article 3). We can see the possibility of the emergence of hybrid commercial arrangements that pave the way towards the institution of heritage commons. But at the same time, it must be recognized that the possibility of using royalties and thus privatizing this temporarily digitized heritage can be seen as an undeniable obstacle, at least, regarding the construction of such commons. This raises the thorny issue of the financing of this digitization process, which seems to us to be at the heart of this issue.

## 3.2. The production methods of heritage commons

### 3.2.1. *The Google challenge*

In the scientific editorial ecosystem, we have shown that the first significant initiatives to create a digital archive were initiated by isolated researchers, like Paul Ginsparg with the arXiv project, and then others, like HAL or @rchivSic. In the context of the digitization of book and cultural heritage, the pioneering Gutenberg project is also the result of a citizen initiative with no connection to the world of library professionals. It was only later, again, that for library professionals, the construction of a digital library became a challenge to be taken up. The arrival of Google with its project to digitize, on a massive scale, the collections of libraries in the United States, and then elsewhere in the world at a later date, has certainly raised concerns. While this Google Books project[15] has offered a new perspective of making the book wealth of the largest libraries accessible at the click of a mouse for all Internet users, and of promoting access to information, it has been the subject of strong criticism, in the United States as on the other side of the pond, by library professionals.

Robert Darnton (2009), historian of American books and director at that time of the Harvard Library (2007–2015), was one of the spearheads of this

---

15 Google Books is in fact a platform hosting a database with an internal engine. In terms of use, the Internet user can either go to the platform's site and search directly, if he or she is only looking for content from books, or use the Google engine, where he or she will be able to access results composed of both web pages and excerpts from certain relevant books. By 2012, 20 million books were accessible through this project. Google's goal at that time was to digitize 15 million books in 10 years at an estimated cost of between $150 and $200 million.

battle across the Atlantic. According to him, Google's monopoly on the process of mass digitization gave it a dominant position on documentary prescription and on the availability of works in the public domain. From this point of view, this project constituted a real threat, which all the more removed the hope of seeing the birth of what he called "a digital knowledge republic" inherited from the Republic of Letters of the 18th century, in reference to the institutional origin of literary property, which was to put the public good before private interest. Robert Darnton particularly blamed the American government, which should have initiated such a project. For him, it is now too late, the battle was lost in advance, as this passage shows:

> Looking back at the digitization process since the early 1990s, we can see now that we have missed a great opportunity. A project of Congress and the Library of Congress, or a grand alliance of research libraries supported by a consortium of foundations, could have done this work at a realistic cost and with a design that would have put the public interest first. We could have created a National Digital Library, the 21st century equivalent of the Library of Alexandria. Now it's too late. Not only have we failed to realize this possibility, but, what is even worse, we are allowing a public policy issue – the control of access to information – to be settled through a private trial (Darnton 2009, author's translation).

This passage is very instructive because it reveals the intrusion and domination of private companies in a sector hitherto reserved for non-market cultural institutions, libraries.

A little earlier, in France, similar concerns were expressed by the then president of the BnF (Bibliothèque nationale de France), Jean-Noël Jeanneney (2009), in a book with the eloquent title: *Quand Google défie l'Europe : plaidoyer pour un sursaut*. However, far from drawing a fatalistic conclusion, this intrusion was for him a "stimulating shock" that should, on the contrary, justify the implementation of a large-scale digitization project on a European scale, the only way to fight against a new form of what he called American cultural imperialism. The digitization projects initiated by the BnF and the Europeana project are part of the extension of this call, which we will study more closely. Since then, a certain number of these cultural institutions have taken the measure of the challenge and have in turn embarked on mass digitization projects or smaller-scale projects oriented towards the constitution of specific digitized collections.

Upstream of these criticisms, an issue emerges that is just as important as the one mentioned above concerning the modalities of implementing a digital library in this new documentary order that the digital ecosystem constitutes. This is the financial challenge directly associated with any project for the mass digitization of cultural heritage. It is often overlooked that such a project is akin to a genuine process of producing a new, non-rivaling and non-exclusive cultural asset, in other words a genuine digital public good. The assumption of this production process by the commercial sector, following the example of Google, makes available to all an unequalled offer in quantitative terms, and also poses real threats to the preservation of the public domain and the visibility of cultural heritage in all its diversity. Conversely, citizen initiative projects are smaller in scope and visibility, but much more respectful of the public domain. We would now like to look at the cultural heritage digitization projects initiated by library professionals and assess the extent to which they lead to the creation of new cultural commons.

Cultural institutions such as libraries are not intended to make a profit; the challenge is therefore to find a viable financing solution for this new type of cultural asset that is compatible with the possibility of access to the greatest number of people, a source of social value creation. It might seem relevant for the State to take charge of the production of this type of good given the positive externalities it generates in terms of education, research and collective well-being as a whole. The argument of the multiplier effect of public investment can also be invoked. However, the cost of producing digitized cultural goods on a large scale is so significant[16] in a context of reduced and controlled public spending that public authorities have encouraged the choice of public/private partnerships. Thus, libraries are faced with the need to seek alternative and complementary financial resources to finance their digitization activity. Their challenge is to find forms of financing that complement or replace public subsidies and that

---

16 We can take the examples given by the Teissier report: in the 1990s, the US Library of Congress developed an ambitious digital policy that led to the digitization of more than 5 million open access documents, at a cost of approximately $45 million. In 2009, Japan began its own publicly funded digitization program for its national library, with an investment estimated at around €90 million for 2010 and around €1 billion for the entire program.

reconcile an effective form of return on investment with the free reuse of digitized public domain works. Several solutions are available to them. They may consider finding sponsors or commercial partners. The latter solution has been preferred in the context of large-scale digitization projects. It has often been criticized because of the risks of enclosing public domain works, with a possible restriction of access rights and a ban on commercial reuse.

To assess these risks and see whether, in the end, these partnerships harm or contribute to the emergence of heritage commons, we propose to study two exemplary examples in France of such partnership projects with one or more private actors: the partnership between the Bibliothèque municipale de Lyon (BmL) and Google, and the partnership between the Bibliothèque nationale de France (BnF) and three commercial actors (Proquest, Believe Digital and Immanens). In each case, we propose to study the conditions of their eligibility for the status of heritage commons.

## 3.2.2. Public/private partnerships: threat or opportunity?

### 3.2.2.1. The partnership between the Bibliothèque municipale de Lyon and Google Books

This 10-year partnership, signed in 2008 between the Bibliothèque municipale de Lyon (BmL) and Google Books, aimed to digitize approximately 450,000 royalty-free heritage works (dating from the end of the 16th century to the beginning of the 19th century). The BmL is the first library in France and, to our knowledge, the only one to date to have signed an agreement with Google for the digitization of its collection. The digitization project, valued at approximately 60 million euros, is without counterpart in terms of financial investment for the BmL. BmL received a copy of the digitized files and in exchange granted Google a 25-year commercial exclusivity. The royalty-free digitized works remain freely accessible, free of charge for Internet users, and can be consulted simultaneously on Google's search engine and on the Bibliothèque de Lyon's website (which was activated in 2012 under the name Numelyo[17]).

For Google, the counterpart of this massive investment is the possibility to accumulate words to index (here in French) to create referencing and sell advertising, in other words, to improve again and again the quality of its

---

17 https://numelyo.bm-lyon.fr/.

search engine. At the beginning, Google had an exclusive right to index digital content, but gave it up quite quickly in 2010[18].

It is interesting to note that the terms of the contract have evolved over time. In 2016, the conditions of use of the digitized content on the platform were drastically modified, as noted by the curator and defender of the commons Lionel Maurel[19]. Initially, the licenses used to protect the cultural heritage digitized on Numelyo were Creative Commons licenses that were not very permissive (CC BY NC ND), prohibiting any commercial use (NC) and any modification (ND). Now, this heritage is placed under the Etalab Open License. As Lionel Maurel notes, this is an evolution that goes in the direction of a distribution policy more respectful of the public domain. Indeed, he rightly points out that the use of a CC license for a work in the public domain is similar to a practice known as copyfraud, which raises a problem of legal validity, since only the owner of a right can make use of it. However, in the case of faithful reproductions of two-dimensional works, the condition of originality indispensable to claim a copyright is lacking in this particular case, which deprives the CC license chosen of its basis.

If we consider this evolution at a legal level, we can deduce that this new situation, conferring identical rights of use and reappropriation to all Internet users, constitutes a favorable condition for the constitution of digital heritage commons. If we compare with the initial situation, we can also see that these rights have been extended because, previously, (physical) books in the public domain were only accessible to registered users of the BmL, with most likely even stronger restrictions for fragile ancient books. The current situation thus reinforces the mission of the BmL to promote the widest possible access to the digitized cultural heritage.

### 3.2.2.2. *Partnerships between the Bibliothèque nationale de France (BnF) and market players*

The BnF embarked very early on with a document digitization program with the Gallica project (1997). But this process was out of all proportion to Google's resources. Until 2007, the volume of digitization was 5,000 documents per year, rising to 100,000 in 2009. From the 2010s, with the

---

18 www.actualitte.com/article/reportages/google-books-numerisation-illegale-du-patri cultural-monk-a-lyon/58084.

19 https://scinfolex.com/2016/08/23/la-bibliotheque-de-lyon-libere-le-domaine-public-en-passant-a-la-licence-ouverte/.

arrival of its new director, Bruno Racine, and under the impetus of the new government, several mass digitization programs were initiated as part of the "Great Loan", which provided a specific budget for the digitization of cultural heritage with the project to change scale. The number of documents accessible in Gallica at the end of 2018 reached nearly 5 million, compared with 4.3 million at the end of 2017. Of this number, nearly 4.2 million were from the library's collections and nearly 800,000 from partners' collections, either directly available in Gallica or indexed[20] only.

Two objectives were targeted by this program, as recalled in the Teissier report[21]: to promote an exhaustive digitization of royalty-free and copyrighted works while avoiding a segmentation of the heritage, and to make this French heritage visible on the Web. Indeed, it was found that French royalty-free works were not very visible via Gallica, but were nevertheless very quickly accessible on the Web via the digitized collections of American libraries. With this in mind, cultural institutions, with the BnF at the forefront, have been encouraged to establish partnerships with private actors (publishers, search engines, distribution platforms) on a basis that is beneficial to all parties (in particular by avoiding exclusivity clauses in the public domain). It is interesting to note that the Teissier report advocated close cooperation between libraries at the territorial level to increase the efficiency of the digitization process, thanks to common digitization centers that harmonize methods and files (on metadata, formats and indexing methods) so as to make them interoperable, as well as the pooling of their expertise and resources, and a common storage center, such as Gallica, that would host the content, which would also remain searchable in their specific environment.

The BnF has entered into several types of partnerships with private actors that are based on different implementation modalities.

As a first step, it launched a call for projects in 2011 for two digitization projects. Two companies were selected[22]. Believe Digital, the European leader in digital music distribution, was selected for the digitization of all of

---

20 See the BnF activity report, available at www.bnf.fr/fr/ra2018-gallica-et-la-politique-de-diffusion-numerique-des-collections.

21 Teissier, M. (2010). La numérisation du patrimoine écrit. Report, ministère de la Culture et de la Communication, 21.

22 www.lemonde.fr/idees/article/2013/02/01/la-bibliotheque-de-france-au-defi-de-la-numerisation_1826114_3232.html.

BnF's sound collections dating from before 1962. This concerns 700,000 titles. Digitization is planned for a period of seven years with, in return, a commercial exclusivity of 10 years from the beginning of the process. Since a large number of the titles are under copyright, it will be possible to listen to excerpts on Gallica; Believe Digital will distribute them via multiple platforms such as Deezer.

The second agreement concerns the American publishing company Proquest. It is based on the digitization of 70,000 old books (most of them in Latin) by this company. The digitization period is six years. The commercial exclusivity on the database is 10 years. In return, Proquest sells access to its database for foreign countries in the form of subscriptions offered to universities. At the BnF, access to the entire digital corpus is possible within the physical premises, but only for library users (approximately 1,500 daily readers), with researchers as the main target audience. It is also accessible to subscribers to the *Early European Books*[23] program. Once this 10-year period has elapsed, the digitized content will be made freely accessible to all on Gallica. The terms and conditions of use are subject to its general terms and conditions of use[24]. In particular, non-commercial reuse is free and unrestricted, with the sole constraint of respecting the attribution of authorship. However, no specific license is affixed to the digitized books, making it easy to identify the possible uses associated with them. It is just called "public domain". As for commercial reuse (the resale of content in the form of elaborated products or the provision of services or any other reuse of content that directly generates revenue), it is subject to a fee and a specific license. We are here within the framework of the application of the Valter law.

Part of the financing was public via the *programme des investissements d'avenir* (future investments program) (2013), for an amount of 5 million euros. The other part was financed by the private company. Bruno Racine explains that this digitization process is accompanied by significant scientific work of data description. Such short-term restrictions were justified by the speed of the digitization process, which according to him would have taken more than 25 years if it had been financed with public funds. He also points out that this led to the creation of more than 40 jobs in France. Faced with

23 www.proquest.com/products-services/databases/eeb.html.

24 https://gallica.bnf.fr/html/und/conditions-dutilisation-des-contenus-de-gallica.

the many criticisms that have been made, particularly by the library world, evoking a "privatization of the public domain", Racine specified that the Proquest company does not have a commercial exclusivity on works belonging to the public domain; any publishing company wishing to digitize them is authorized to do so. He was very critical of the demand for "immediate free access", which could slow down the digitization process[25].

The BnF has also entered into another type of partnership with the company Immanens leading to the creation of the Retronews portal, distributing more than 400 digitized press titles. The BnF is in charge of selecting the titles to be digitized in conjunction with the curators, overseeing the digitization operations (handled by an external operator, Adoc Solutions), designing the editorial content and managing the development and operation of the site. Immanens manages the search tool and page indexing and has developed a dedicated viewer.

In this case, high value-added paid services are offered to users. The distribution of the digitized documents is free and open access online. But in addition, this portal offers advanced search tools specific to press collections, as well as additional editorial content for a targeted audience (students, researchers, journalists, etc.). To access it, the user must pay a subscription. This is an economic model of the freemium type. This agreement was reached through its subsidiary BnF-Partenariats, which covered part of the cost of the digitization process (part of the archives was already available on Gallica).

### 3.2.3. *On-demand digitization and citizen contribution*

For digitization projects that are more limited in terms of volume, some libraries opt for participatory citizen funding via a call for public "sponsorship" in various forms. The two examples on which we rely have been studied beforehand by Lionel Maurel[26]. Here we take up some elements of his analysis.

---

25 www.lemonde.fr/idees/article/2013/02/01/la-bibliotheque-de-france-au-defi-de-la-numerisation_1826114_3232.html.

26 Maurel, L. (2016). Quel modèle économique pour une numérisation patrimoniale respectueuse du domaine public ? [Online]. Available at: https://hal-univ-paris10.archives-ouvertes.fr/hal-01528096/document.

In addition to its partnerships aimed at so-called mass digitization, the BnF has also set up a sponsorship system, under the name *"Adoptez un livre"* (adopt a book)[27], based on a crowdfunding logic: users choose a document that they wish to to be digitized from a catalog previously developed by the BnF. This donation is 66% tax deductible. The digitized document is put online on Gallica with the donor's name. These books are placed under Gallica's general terms and conditions of use with a restriction on the commercial use of files subject to prior authorization and payment of a fee. Consultation of the database of books digitized by these "citizen" funding sources reveals that only 400 books are currently digitized.

Another project, called Numalire, is based on a similar principle while targeting the book heritage of several libraries. It is an experiment in digitization on demand of the heritage preserved by libraries in the form of crowdfunding[28]. In 2013, YABé launched an eight-month experiment aimed at digitizing and republishing on demand royalty-free documents held in eight Parisian libraries, with the aim of sharing the costs of digitization with Internet users. The company was remunerated on the sale of paper books on demand. The digitized documents were in the public domain. YABé created a website with the records of 500,000 documents. An Internet user interested in a book could request a quote and the company would forward the request to the library concerned, then open a financing subscription on its site if the project was deemed relevant. The Internet user was then invited to distribute this subscription via social–digital networks. Once the financing was reached, the company launched the digitization via an external service provider. At the end of the process, a file was given to each subscriber and the library. During the experiment, 414 requests for quotes were made, but only 11% were processed (in many cases, the book could not be digitized). In the end, only 36 digitizations were carried out. The digitized books were made available under a PDM license. This project has not been generalized.

### 3.2.4. *The heritage commons: a plasticity of forms*

Digital cultural heritage cannot be categorized as a cultural commons as such. The first clue that led us to its potential eligibility for such a status stems from its specific nature as a public economic good due to its properties

---

27 https://gallica.bnf.fr/html/und/adopter-un-livre-offir-une-voix.

28 For more details, see http://bbf.enssib.fr/contributions/numalire.

of non-rivalry and non-excludability. But the study cannot stop there, because this property is not in itself a sufficient condition. We have therefore shown the interest of paying specific attention to the modalities of implementation of the production process of this type of digital cultural good and, in particular, to locate and identify the legal rules that determine the outlines of the ownership of this digital heritage as well as the preferred economic exploitation model.

Such an investigation revealed the plurality of production processes and, in so doing, revealed distinct rules of shared ownership and the accompanying rights of use and appropriation. These differences reveal how the status of digital heritage commons does not depend on a single model, but can, as in the case of land commons, accommodate different bundles of rights.

In the first two cases studied, the production process of the cultural heritage of libraries (BmL and BnF) was carried out by a third party company from the traditional commercial sector (Google, etc.). The digital cultural resource produced was not the exclusive property of the producer. It was owned jointly by the producer, the libraries (which could be referred to as "curators") and a third category of end-users (consumers from an economic perspective). We find here the case of a conception of shared ownership as a bundle of rights. These three categories of actors had different rights of access and appropriation of this resource (operating rights in the sense of Ostrom), and some of them (the first two) had collective regulation rights.

In the case of the BmL–Google partnership, the production process for works in the public domain was handled by a third party, in this case the company Google, which bore the entire cost of digitization. However, this company does not have exclusive ownership of the good produced, which is similar to a database containing digital resources and associated metadata. This unrivaled digital asset has resulted in the production of two copies, one held by Google, the other by the BmL. This is the case of a non-competing cultural asset where ownership is shared. This shared ownership confers different rights on the two partners:

– a right of commercial use and an indexing monopoly for Google only for a given period (25 years initially, reduced to 8 years);

– a dissemination right for the BmL with the possibility of putting in free and open access to the digitized content for Internet users on the Numelyo platform. These same rights evolve over time as the commercial monopoly comes to an end, and users' rights are extended with the Etalab license, which is very permissive and affixed to all content.

We can therefore say that we are in the presence of a genuine bundle of non-exclusive property rights between the various stakeholders: Google, the BmL and the users. While there has been a form of privatization of digital heritage, it is partial and has been restricted in time to ensure that Google can reap the benefits of its initial investment. In this perspective, all the conditions are required to affirm at the end of this study that this project contributes to the constitution of a cultural heritage commons.

In the BnF–Proquest project, the production process is taken care of by the company Proquest. As in the previous case, this shared ownership is accompanied by rights of exploitation of the resource by the merchant company in the form of a commercial exploitation monopoly quite similar to the previous partnership (10-year term). The "curator" (BnF) also benefits from a right of diffusion (access) to users, but not of the same nature as in the previous partnership. On the one hand, the right of access granted by the BnF is more restrictive than in the case of BmL, because it does not give free access to everyone via Gallica once digitization is complete. There is a restriction on this right of access, of a geographical nature, only to BnF users via the computers of this library, during the period of commercial operation. On the other hand, once the period of commercial exploitation has expired, it can be seen that uses in terms of appropriation and reuse are also less permissive, since the commercial uses authorized by the BnF are subject to a fee (and therefore subject to payment). The BnF has granted itself the right to levy a fee (without any time limit) in the event of a request for commercial use by a third party. This form of commercial exploitation is not aimed at profit-seeking as in the case of commercial enterprise (Proquest).

However, it can be seen that the right to levy a royalty is not limited in time. Does this last clause constitute an obstacle to the eligibility of this digitization project for the status of heritage commons, which, moreover, presents important assets to claim it? We do not think so. Nevertheless, it does show that there is not just one generic model of heritage commons, but several, each characterized by clusters of rights defining the singular shared property. On the contrary, it is indisputable that some are more respectful of the digital public domain than others.

In this respect, there are two other cases that also deserve special attention: the case of the partnership between the BnF and the company Immanens and the case of digitization projects based on a crowdsourcing logic.

In the first case, the production process belonging to the digital cultural resource is not entirely outsourced by the "curator" to the commercial enterprise. It is the subject of joint production and cost sharing between the two entities. All users have a right to free access and copy these digitized documents. On the contrary, digitized production is divided into two categories of offers according to the logic of the freemium economic model: a raw offer open to all and an enriched offer that is the subject of commercial exploitation by the two entities producing the resource. Here we are confronted with a third form of heritage commons based on a set of singular rights.

In the second case, two scenarios have been evoked. In BnF's "Adopt a Book" project, the production of digitized cultural heritage is based on external funding supported by citizen contributors. The BnF obtains a distribution right by giving all users access to these resources via its Gallica platform. Contributing citizens obtain a form of recognition for their financial participation (by affixing their name as a donor to the digitized work) and indirect financial compensation (in the form of a tax credit). They have the same right of use as any other user. This right of use is not very permissive, since any commercial use implies the payment of a royalty. Finally, in the case of the Numalire project, the production process is shared between a commercial enterprise and contributing citizens. The former has the possibility of commercially exploiting printed copies of digital files. Ownership of digital cultural resources is shared between the company, the library (which receives a digitized file of each work from its collections) and the end-users, who have a very permissive right to access and reuse them (Etalab public domain trademark license). Here again, we are faced with an entirely singular situation in terms of shared ownership and bundles of rights.

Along the way, this study has allowed us to show how these different digitization projects each contribute to giving life to heritage commons that share design principles, but are distinguished by the nature of the bundles of rights and the economic exploitation of the digitized cultural heritage. It now remains to study a fundamental dimension that constitutes a knowledge commons: its enrichment.

## 3.3. Governance issue: enriching our common heritage

A digital heritage commons is one of the manifestations of a knowledge commons whose social value is directly correlated with its capacity to be used, exploited, shared and disseminated. But what does this enrichment mean in the context of building a digital library? At this stage, the existing literature does not give us a precise answer, except to mention that this is a governance issue. In our opinion, the question of enriching a digital heritage commons can be considered along two distinct paths. Upstream, at the level of information infrastructures, the challenge is to encourage the construction of a shared heritage space, an indispensable condition for giving the opportunity for possible expansion of the shared heritage. Downstream, it is a question of implementing conditions that are favorable to its appropriation and exploitation, a challenge that involves not only a policy of making this digital heritage visible, but also a policy of editorializing content to broaden the base of the public likely to make use of it.

### 3.3.1. *The construction of a shared heritage infrastructure*

At the production level, the challenge for the governance of a digital heritage commons is to encourage the construction of a shared heritage space. Indeed, such a shared space is not intended to be enclosed and circumscribed within an institutional perimeter, even if it is at this level that it takes shape. Rather, its vocation is to grow and broaden its outlines by gradually bringing together the different digitized productions, here and there, of digitized heritage.

Thus, the construction of a digital heritage commons cannot be envisaged without a desire for cooperation between cultural institutions, in this case libraries, leading to consultation and agreement on two types of rules: legal rules relating to the methods of use and reuse of digitized cultural heritage, and technical rules relating to the development of a shared space for the metadata associated with it. Its sustainability is at stake. Indeed, without the harmonization of these rules, the growth of the shared space is necessarily affected.

We have studied at length the role played by the existence of institutional rules in terms of access and reuse of data that are as permissive as possible. Another type of rule plays an equally important but rarely emphasized function in the commons literature: the rules that prevail in the metadata construction strategy. Hess and Ostrom did not insist on this dimension in their study of scientific digital archives. However, it deserves special attention, because without a coordinated strategy at the level of data interoperability, the social utility of such a digital commons will necessarily be reduced.

In the field of digitization of museum content, researchers have been working on this issue. For example, Brigitte Juanals and Jean-Luc Minel (2016) devote part of their work to this essential question of building a shared heritage space in the Open Data Web by questioning possible forms of interoperability. The existence of an Open Data Web ecosystem should encourage the institution to move away from a reflection centered on its resources and thesauri, to leave a model of information organization in the form of local databases, to integrate an inter-institutional dimension that can lead to the creation of an open and shared digital cultural space based on standardized description languages of the Semantic Web and information access (such as the Europeana EDM standard that we will discuss later). This challenge of technical standardization is, in the reality of the heritage world, far from being met. However, "from a technical point of view, redocumentarization implies the adoption of the founding concepts of the Semantic Web and the constraints inherent in the formalisms of the technologies and description languages associated with them. However, for heritage institutions, it is essential to emphasize that this technical reconfiguration is inseparable from an internal reorganization and an opening of their resources within the open data Web" (Juanals and Minel 2016, p. 5, author's translation). Based on an empirical study conducted in connection with various cultural heritage institutions, Juanals and Minel show that this process of adopting international technical standards runs counter to their practices, which often consist of adopting local and disciplinary thesaurus standards. This is particularly the case for museums whose collections are composed of unique objects. For libraries, on the contrary, they point out that these institutions have long since adopted classification systems that correspond to international documentary models.

The production of digitized content that is freely accessible and open to reuse is only one of the conditions that must prevail in the construction of an authentic common heritage in the public domain. Indeed, the construction of such a commons implies questioning the relevant level or scale of its construction. Even if the BnF very probably contains the most important pool of books belonging to the public domain in France, other resources can also be found elsewhere, in other places. Moreover, it has been noted, in the few examples mentioned, that there is no prescriptive cultural policy at the territorial level regarding the rules to be followed in terms of a digitization strategy. However, we may wonder whether it is relevant for each cultural institution to carry out a digitization strategy independently and uncoordinated, knowing the cost that this represents. This is why it seems to us that the elaboration of such commonalities has real relevance only at a meta-institutional level, which should result from collective and concerted action between different cultural institutions in conjunction with the public authorities. The Europeana platform is a very interesting example to mention, as it is an attempt to pool the whole of Europe's digitized cultural heritage.

Europeana's governance reveals a very democratic structure with representation of all project stakeholders, from the founding members through the partner institutions to the reusers. Europeana has proposed the implementation of a voting system for all decisions proposed by the governing board. It established a "Public Domain Charter"[29] aimed at all potential partners, specifying that "the digitization of public domain content does not create new rights in it: works that are in the public domain in their analog form remain in the public domain once they have been digitized"[30]. Thus, it encourages them to use PDM licenses for public domain content and CC licenses for others.

At the level of information infrastructure, Europeana is pursuing a proactive policy to promote international harmonization at the level of metadata. Europeana does not archive works, but only serves as a research interface by publishing metadata from partner institutions. It harvests the metadata of the partner institutions and provides a link to their site for the effective consultation of files (through the OIA-PMH protocol). To promote the emergence of a technical standardization of this heritage space, it has published the "Europeana Publishing Guide"[31], which defines minimum

---

29 This charter, published in 2010, is available online: https://pro.europeana.eu/files/Europeana_ Professional/Publications/Public_Domain_Charter/Public%20Domain%20Charter%20-%20EN. pdf.

30 *Ibid*, p. 2.

31 https://pro.europeana.eu/post/publication-policy.

rules to be respected by partner institutions that want to share their data and cultural content.

The quality of the metadata is considered an essential dimension on which the relevance of the platform is based: "The quality of your data is really important because in a database as big as Europeana records can get lost easily. It is therefore essential that the metadata have enough useful elements to make the content findable."

It is therefore essential that the metadata contain enough useful elements to ensure that the content can be found[32]. Quality data provides users with a better user experience and a quality connection to partner institutions' collections. Specifically, metadata must comply with the Europeana Data Model (EDM) in order to be validated (self-verification process).

Despite this strong willingness of the Europeana platform to establish a set of common rules of a legal and technical nature, this harmonization, which is indispensable for the enrichment of a digital heritage commons, still faces many obstacles that should be studied more carefully, to assess whether such a project is doomed to failure or whether it just needs time for cognitive and cultural learning on the part of the various stakeholders.

### 3.3.2. *Content editorialization and digital mediation*

Promoting the widest possible access to digitized content: this is the objective that libraries that have embarked on a digitization strategy must pursue. But as the Teissier report (p. 27) repeatedly stresses, "there is no point in being available if you are not visible". In other words, any digitization strategy must also be accompanied by a strategy for making content visible beyond the platforms that host it on the Web. In the cases studied above, we will show that such a need has been taken into account and that it has been translated into editorial actions (in the sense of the process of knowledge production in the digital age) of this content, placing the user no longer as a simple "consumer" of digitized cultural resources, but as a "contributor" likely to create new resources in turn. In this sense, the governance of digital cultural heritage is clearly oriented towards its permanent enrichment.

---

32 https://pro.europeana.eu/files/Europeana _Professional/Publications/Europeana%20Content %20Strategy.pdf, p. 9.

All the cultural institutions that have initiated a digitization program have gradually put in place an editorialization strategy to facilitate the ownership of content and make it accessible to a wider audience. If its social value is to be increased, the digital heritage commons must be appropriated beyond the primary users, such as researchers, the educational world or the cultural institutions themselves. Such a commons must be addressed to the general public in order to increase its social usefulness. Now, to facilitate this dynamic of appropriation by the general public, institutions have set up digital mediation actions that position the first "users", who could be described as "early adopters", as contributors of new digital cultural resources. This is a path that has been taken by many institutions in a complementary way to their digitization. Let us give a few examples.

The BnF has embarked on this type of operation at several levels. First of all, it created Gallica Studio, offering, according to Olivier Jacquot, research coordinator, "a space that is at once a playground, a toolbox, and a showcase for innovative and creative reuse of the content available on Gallica". This dedicated editorial project revolves around three types of space: a "toolbox" offering tutorials, an iconographic search engine, APIs, and data sets. Gallica Studio then supports the implementation of collaborative projects such as *hackhaton*. In 2018, for example, the "*Mix tes romans*" (mix your novels) project was designed for high school teachers and their students. "Using the application and a custom-generated deck of cards, students use resources from Gallica, wikidata, and Radio France broadcasts and podcasts. They use these resources to create other stories and share them with the #MixTesRomans community"[33]. Finally, contributors, known as Gallicanautes, are Gallica users who contribute to the dissemination of Gallica documents on the Web via their blog or personal, associative or institutional site. Through their contributions, they participate in building the identity of the digital library.

At the scale of Retronews, the BnF has also implemented an editorialization strategy[34] aimed at transforming the archive into edited content. For example, the site features contributions from journalists, researchers and historians who, in a logic of different forms of media production ("echoes" daily papers, "chronicles" or "retrospective cycles"),

33 http://gallicastudio.bnf.fr/Hackathon2018.

34 Manchette, É., Thouny, N. (2018). RetroNews [Online]. *Bulletin des bibliothèques de France (BBF)*, 15, 32–35. Available at: http://bbf.enssib.fr/consulter/bbf-2018-15-0032-004.

place the articles in their political, social and media context. The BnF has also opened up a contributory logic (in the form of comments) and the content is broadcast on all screens and the main digital social network platforms (Twitter, Facebook, YouTube). The objective for the BnF is to move from a consultation logic to a logic of a site that "emits" content.

If we look at the BmL, we can note a similar editorialization strategy that accompanies the visibility of digital archives. Gilles Eboli, director since 2011, explains that, as in the material universe, the challenge is not only to build collections, but also to make them accessible to a wide audience for a true sharing of knowledge through digital mediation operations. For example, in the field of heritage enhancement, "the objective will be to multiply the proposals that place the public in the position of actor, to diversify the angles of reading and appropriation through a mixture of forms of display and restitution"[35].

Finally, the Europeana platform has also created a site dedicated to potential reusers of open data: Europeana Pro[36]. It is aimed at four specific targets: the research world, the creative industries, the depositary institutions of cultural heritage and the educational world. This platform implements an editorial strategy to make the content of the site more attractive to the general public, and also to encourage the reappropriation of the proposed content, whether from a commercial or non-commercial perspective. Europeana Labs is the entity created to foster the implementation of this type of project: "Europeana Labs is the ideal place for those who have the imagination, skills and desire to play with digital cultural content and use it in their experimental work or sustainable commercial projects."[37]

Europeana's strategy for reusing digital cultural heritage is very clear and explicit on their website. It is primarily aimed at "reusers", that is, professionals in the education, research and creative industries sectors. Europeana will accompany them in their creative and innovative projects by providing them with free resources. It is therefore also and above all the users who will also participate in the enrichment of the digital heritage commons.

---

35 Eboli, G. (2018). Stratégie éditoriale de la BM de Lyon [Online]. *Bulletin des bibliothèques de France (BBF)*, 15, 36–45. Available at: http://bbf.enssib.fr/consulter/bbf-2018-15-0036-005.

36 https://pro.europeana.eu/home.

37 https://pro.europeana.eu/what-we-do/creative-industries.

To encourage these dynamics, Europeana offers free resources (inputs) to the different types of potential reusers. It supervises and supports the implementation of crowdfunding campaigns for creative reuse projects in the field of education in partnership with Gote[38]. Here is the presentation of the latest @edTech Challenge, aimed at fostering creative and innovative projects articulating culture, education and technology:

> Today we launch the **Europeana #edTech Challenge**: a funding opportunity for entrepreneurs, developers, designers and educators who explore digital opportunities for education and training. Do you have an exciting project in the intersection of **culture, education and technology**? We're offering €30,000 for the best products, services or businesses that bring together cultural heritage and educational technology[39].

Finally, Europeana organizes online competitions (with a monetary reward) to select the best ideas for a creative reuse of digital cultural heritage in a crowdsourcing format. It also offers APIs that give the possibility to build innovative applications.

Thus, in addition to the strategy of digitizing works that have fallen into the public domain, leading to the production of new digital cultural goods that can be likened to an institutionalization of shared heritage, the cultural institutions studied have implemented various strategies for the editorialization of this same content in order to make it accessible to a wider public. We have also noted that users are encouraged not to remain confined to their status as "readers" by actively participating in the construction of this new editorial edifice. It is based on a participative logic that is part of free culture by encouraging the development of creative projects, whether in research and education or amateur culture, or for innovative projects.

---

38 http://goteo.org.
39 https://pro.europeana.eu/post/europeana-edtech-challenge.

# 4

# The Written Commons in the Publishing Industry

The written commons are identifiable in the form of digital archives of written content, stored and made visible on platforms. They offer amateur writers new formats of writing practices associated with offers of intermediation services.

The rise of this type of platform is explained by some of the major transformations that the publishing industry is undergoing, as we will see in section 4.1.

However, not all of these platforms can claim such a written commons status, in the same way that not all amateur creative practices are part of free culture. Identifying the written commons in the digital book ecosystem involves turning our attention to what constitutes their purpose: the desire to foster the development of free culture promoted by the multitude of creative practices of amateur Internet users in the field of the written word.

Based on this preliminary observation, we have selected several platforms because of their potential eligibility for commons status, identifiable by the fact that they offer Internet users the possibility of sharing their digitized written content using a Creative Commons license.

In each of the cases studied, we will seek not only to rule on the eligibility of such platforms for written commons status, but also to study ways of cohabitation with the actors of the digital publishing sector.

## 4.1. The transformations of the editorial ecosystem

### 4.1.1. *Digital textuality and new uses*

Roger Chartier makes a major contribution to our understanding of the transformations of the book in the digital age. According to him, we are faced with a radical change in the way the written word is produced, transmitted and received in the digital ecosystem: "The revolution of our present is obviously more than Gutenberg's; it does not only modify the technique of text reproduction, but also the very structures and forms of the medium that communicates it to its readers."[1] The printed book can be considered an "heir" to the manuscript book, because the criteria for identifying its form and organization are similar (organization in notebooks, the presence of indexes, a hierarchy of formats, etc.). On the contrary, by replacing the codex, the screen, as a new medium for writing and reading, leads to the explosion of the usual categorizations of the printed book: its modes of structuring, organization and consultation. The same medium, the computer screen, makes different types of texts appear that were previously distributed in distinct objects. The same surface, the screen in the electronic world, makes it possible to read all texts, whatever their genre and function: "In the world of digital textuality, discourses are no longer inscribed in objects that allow them to be classified, hierarchized, and recognize their own identity." Digital textuality implies a dissociation from what constituted the foundation of printed textuality: the solidarity between the medium and its inscription[2].

But this is not the only transformation characterizing what is now called the digital book. In the digital ecosystem, the book becomes rewritable by essence and not by accident, as Pierre Mounier points out: "Reading books has been the subject of commentary and discussion for a long time, but until now, the commentary was not visible on the book itself: the traces produced by social uses were disconnected from the printed book, which is not rewritable" (Mounier 2012, p. 34, author's translation). This new property is translated through two novel dimensions: the book becomes a digital file in the form of a set of instructions for software, which can be subjected to a

---

1 Excerpt from an article entitled "Du codex à l'écran, les trajectoires de l'écrit" published in the magazine *Solaris* in 1994, available at http://gabriel.gallezot.free.fr/Solaris/d01/. 1chartier.html.

2 *Ibid.*

computational logic; the book becomes reticular because, like any other digital document, it weaves links with its environment composed of heterogeneous contents through the construction of an intertextuality.

From these transformations affecting the very essence of the book in the digital ecosystem, a much more complex relationship to reading will result. Two major evolutions can be noted. First, as the digital ecosystem resembles a new form of public space, reading is no longer solitary and silent, but open and shared. As Milad Doueihi states, reading has become multiple by now taking shape "on the public sphere of the digital agora, in the form of podcasts, but also in the form of an exchange of fragments, snippets and quotations or the communication of hyperlinks, images and videos... It is a shared reading, visible to all and subject to the recovery and exchange characteristic of the new digital sociability" (Doueihi 2008, p. 309, author's translation). Finally, while the codex, in the form of a printed object, imposes its structure on the reader, who cannot participate in it at all or only in a clandestine and marginal way, electronic text allows the reader to undertake multiple operations: indexing, copying, dividing, recomposing, moving, etc. (Doueihi 2008, p. 309). The latter is thus encouraged to take part in the very evolution of the content produced, thereby blurring the rigid boundary between writing and reading, between the author of the text and the reader of the book. This new possibility calls into question the usual categories used to describe works which, since the 18th century, have been identified with an individual, singular and original act of creation on which literary property law is based. Milad Doueihi sees this as an unprecedented opportunity for the emergence of new forms of original writing freed from the constraints imposed by both the morphology of the codex and the legal regime of copyright: polyphonic, open and malleable, infinite and moving writings. This thus introduces a new reader relationship with the work: "The reader becomes an author not by eliminating the trace of the original creator, but rather by moving the chosen piece, by finding a new context for it, by circulating it in the vicinity of other objects" (Doueihi 2008, p. 308, author's translation).

These new properties linked to the emergence of digital text will induce important transformations in the book ecosystem, which are not without consequences on the traditional logics on which the commercial arrangements are based. Before exploring these different manifestations, another important dimension, related to the transformations of the book, concerns the radical evolutions induced in the documentary order and, in particular, the possibility of creating a universal library. Since the fundamental logic of the library is sharing, the possibility offered by the Web of broadening access to documents places it at the heart of the issues surrounding the constitution of common heritage.

By contributing to the transformation of the book object, the digital ecosystem will bring about major transformations in the traditional commercial arrangements of publishing and libraries. Highlighting the most significant features of these transformations will help us to better understand how so-called free cultural practices contribute to the evolution of these arrangements and to what extent they condition the emergence of cultural commons. Since the 19th century, the publishing industry has been structured around the central figure of the publisher, who plays a central intermediation function between authors upstream and distributors and readers downstream. The editorial model based on the legal institution of copyright has become the dominant socio-economic model of this sector. Two major changes have shaken this established order. First, in the form of a computational and reticular file, the digital book will constitute a new form of immaterial physical capital for the commercial actors of the attention economy who now dominate the new documentary order. Second, the development of contributory creative practices on the Web will also contribute to the destabilization of traditional actors in the publishing world. Let us consider these two points in turn.

### 4.1.2. *The digital book immersed in an attention economy*

Since the end of the 1990s, the cultural book industry has seen the emergence of new players, most notably Amazon, Google and Apple. By taking a downstream position, on the distribution side, they have rapidly captured a significant part of the value of the sector by organizing access to books (paper and digital). They have also disrupted the existing logic structuring this value chain by making the logic of the digital economy work in their favor, such as the network effects linked to increasing returns on adoption[3] (Benhamou 2014). Thus, the strategies of these players consist of first reaching an optimal network size from which the number of users will

---

3 The more the number of Internet users using a service such as the Google search engine increases, the more its value will increase because the satisfaction drawn by an individual depends directly on the number of its users. This principle applies to many other sectors, but finds particular application in the field of the digital economy.

continue to grow. Each one, in its own way, has then implemented various strategies to make these network effects work in their favor. We will give a few examples.

Amazon offered a medium for reading digital books, the Kindle, which is mainly designed to lock the consumer into his or her offer by preventing him or her from reading content purchased on its platform on another reading medium. Then, this company gradually enriched its services with online subscription offers and self-publishing services to build customer loyalty and gradually lock them into its merchant universe. It is also gradually moving up the book value chain by investing in publishing. Finally, it has also introduced the logic of attention with the implementation of a free automatic prescription service as part of its merchant offering.

Apple is doing the same on a larger scale by enclosing the consumer in its own digital ecosystem, with a dedicated hardware offering and a platform for offering all types of content, the Apple Store. As for Google, not only is it the preferred gateway to all forms of content, but it has also made a remarkable entry into the field of books, by aiming to become a key player with its Google Books universal library project.

In all these cases mentioned, while these digital companies are seeking to position themselves in the book value chain, it is above all because it feeds their strategy of creating value from the exploitation of all types of information content (Miège 2007). The digital book thus now competes with all the other products or services that are at the heart of the attention logic. The principle of scarcity, at the very foundation of the market economy, has shifted from the extreme of production to that of reception, giving rise to new socio-economic models: "It is consumers who own the resource that has become the rarest and most precious – their attention – and we can expect to see the generalization of arrangements in which we receive free services (Google, Facebook) in exchange for privileged access to our attentional capacities and dispositions" (Citton 2014, p. 8, author's translation). Henceforth, access to information and communication content is made through search engines that specifically trade in it. For Olivier Ertzscheid, this model works on the basis of an industrialization of indexing in the digital ecosystem: "Our world has always been documented. But, for centuries, indexing, whether human or mechanical, remained out of any commercial consideration" (Ertzscheid 2009, p. 34, author's translation). This commercialization does not imply the setting of a price as in most commercial exchanges. It is based on a model that allows free access to commercialized information, which is otherwise monetized by advertising models (Bomsel 2007). This model of "commercial" free access inspired by two-sided markets, a classic model of advertising financing for the media (television, radio, free newspapers), nevertheless presents a notable difference with the latter, as recalled by Françoise Benhamou and Joëlle Farchy (2014). Indeed, content creators are not remunerated by the actors of the attention economy, unlike the media markets.

From this perspective, for Christian Robin, the stakes are twofold for publishers. Not only is it a question of finding ways to monetize their content in a world dominated by free content, but also of implementing strategies aiming to capture the attention of potential readers: "The book industry, faced with the digital world in the context of an attention economy, forces its players to look at their activity from new angles in order to find the value that alone will allow them to survive" (Robin 2015, p. 50, author's translation).

Faced with these changes, publishers initially adopted a precautionary strategy limiting the digital offer to a small portion until the end of the 2000s: "Their caution is easily explained: fear of digging their own grave when entering a market where the experiences of the press and music do not encourage people to rush" (Benhamou 2014, p. 70, author's translation). Thus, like producers in the music industry, they have sought to protect their profession through a form of conservatism by preserving their traditional socio-economic model (editorial model) through the DRM protection system. They also defended the extension of the Lang law on the price of digital books[4], which reinforced the publisher's strategic function in setting the price and the impossibility for distributors to have discretionary power in this area.

The major publishing houses have also sought to rebuild the traditional book value chain by positioning themselves downstream in the sector, by creating, alone or in partnership, distribution and distribution platforms[5].

4 This is the law of May 26, 2011. It gives the publisher, following the example of the Lang law, the power to set the same selling price for digital books for all resellers, whether they operate from France or abroad. The purpose of this law was to create the conditions for French players to compete on a level playing field.

5 The digital distributor is in charge of the promotion of the digital book catalog, and manages the commercial server hosting the files, the metadata and the sale of the files to

While such a strategy enabled them to maintain their strategic position, they were not content with selling their books on these platforms and had to go through other key intermediaries, such as Amazon, which benefit from increased visibility. In this case, their central position is less assured and negotiations on pricing can be very conflictual, as the example of the conflict between Hachette and Amazon in the United States (Legendre 2019) reminds us.

Significant concentration dynamics have also been observed in some sectors, accompanied by a modification of the economic models in order to maintain their central position. Some publishers have tried to make links with the world of video games, either with a view to transmedia exploitation or with a view to consolidating[6] a globalized communication group.

Since the 1990s, it is above all those involved in scientific publishing who have implemented such concentration strategies with a view to acquiring a dominant position in the digital ecosystem. Through such merger operations, they have been able to rapidly increase the base of their offer, now dematerialized, to their users. This offer, in the form of a giant database of journal articles accessible to their customers, has enabled them to bring network effects into play. Elsevier has become the dominant player in this now oligopolistic market, sharing the market shares with a handful of players such as Springer, Wolters-Kluwer and Wiley. They have also introduced a new, much more remunerative economic model based on a flat-rate licensing system paid for by libraries that allows users[7] to access all the publishers' titles in full text. One of the obvious consequences of this concentration of the scientific publishing market has been, as already mentioned, the increased dependence of university libraries, and therefore universities, on these publishers in the face of an ever-increasing supply.

---

the final customer. A few examples: Eden Livre is a digital diffusion and distribution platform common to the Actes Sud, La Martinière-Le Seuil and Madrigall (Flammarion and Gallimard) groups. Numilog is the platform created by Hachette.

6 See Benghozi, P.J., Chantepie, P. (2017). *Jeux vidéo: l'industrie culturelle du XXIE siècle*. Presses de la Fondation nationale des sciences politiques, Ministère de la Culture et de la Communication, DEPS, Paris.

7 The evolution of the scientific publishing market in the digital ecosystem was studied by Nathalie Pignard Cheynel in her thesis defended at Grenoble 3 in 2004, "La communication des sciences sur Internet: stratégies et pratiques".

### 4.1.3. *The digital book and the growth of self-publishing*

The second change in the book industry is the increasing development of self-publishing[8]. This activity, which has always existed on the fringe of traditional publishing, finds here multiple modes of expression according to the services offered by the platforms that are in this niche. While this activity, which until now has been rather confidential, is now clearly on the rise, this can also be interpreted as the answer to a growing need expressed by content creators. Thus, "one cannot detect only the behavior of the rejected and untalented author", as Françoise Benhamou is inclined to think (2014, p. 137, author's translation). By facilitating the circulation of knowledge and making available to everyone new modes of creative expression, the digital ecosystem has fostered the emergence and deployment of amateur practices in the cultural field (music, sound and image).

This rise of creative practices has been widely discussed in Part 1 on free cultural practices that affect not only writing, but also music, image and video. More recently, CIS researchers have in turn studied this question (Flichy 2010; Proulx *et al.* 2014), seeking to account for the singularity and novelty of these practices. For his part, Patrice Flichy describes this new trend as the symbol of the "amateur's consecration". Seeking to better circumscribe the outlines of the figure of the amateur, he defines it as being "halfway between the ordinary man and the professional, between the layman and the virtuoso, the ignorant and the scholar, the citizen and the politician" (Flichy 2010, p. 11, author's translation). His activity is mainly deployed in the non-market sphere of the arts, public affairs and knowledge. It is rarely a solitary activity and takes shape in collectives, virtual communities within which the amateur can exchange, debate and find an audience. What distinguishes him from the professional is that he is free in the construction of his creative project, motivated most of the time by pleasure, guided by his passions. But this impetus is not incompatible with the search for symbolic, even financial, remuneration. As Cécile Méadel and Nathalie Sonnac point out, in the field of writing, "book writing is almost always a non-professional activity, but the specific characteristic of the Internet is that this means of communication allows non-professionals to find their audience

---

8 Bertrand Legendre points out that the usual distinction between self-publishing, designating the material and/or financial management by an author of the production process of a paper or digital work, and auto-édition, which adds to this process the fact of bringing the work to the public's attention, is more difficult to separate in the digital ecosystem.

without going through intermediaries" (Méadel and Sonnac 2012, p. 112, author's translation).

Blogs and social–digital networks were the first preferred form of expression of this rising wave of amateur culture. For Dominique Cardon, written publication on the Web has become a topic of conversation. The Social Web has made it possible to expose one's "extimacy" in half-light exchanges, in the form of a staging of oneself and thus "democratizing narrative self-construction by inscribing it in the practices of ordinary life" (Cardon 2010, p. 59, author's translation). In the written word digital ecosystem, new forms of writing practices, continuous, open and social, so-called contributory, embody this new trend. Over time, they have gained considerable momentum, gradually blurring the boundary between writing and reading, inserting text into a conversational continuum between author and readers. The emergence of self-publishing platforms, as new forms of intermediation, can therefore be interpreted as a response to these new participatory logics. Their originality is to introduce a selection system *a posteriori* regulated by the crowd of Internet users (through their opinions, comments, etc.). As Bertrand Legendre points out, it is a process of bypassing and erasing instances of legitimization that is at work. Now, the cultural legitimacy held by publishers is no longer a major issue, since the power of selection is now in the hands of the amateur communities of these new platforms. Moreover, these platforms often claim a willingness to democratize freedom of expression: "They do not fail to regularly stress that this decline in mediating functions as a sign of a democratization of the process of access to publication, leaving each author the theoretical possibility of addressing the greatest number of readers via the Web" (Legendre 2019, p. 14, author's translation).

While undeniably responding to an emerging need resulting from the growth of amateur practices in the field of writing, all these platforms are not alike, either in terms of the way they operate or their economic model. Nevertheless, it could be hypothesized that by promoting these creative impulses, freedom of expression, they are *a priori* eligible for the status of new cultural commons in the field of the written word. Wikipedia, as a platform for contributory self-publishing based on the use of Wiki technology, has been endorsed in this sense by various studies aimed at studying its status as a commons. Benkler (2009) had initiated this reflection, as shown above. It was later taken further by other researchers. In France, we can cite the pioneering work of Dominique Cardon and Julien Levrel

(2009), who helped to shed light on this question by studying with great precision the mechanisms at work in the model of self-organization, combining critical vigilance, conflict regulation and graduated sanctions. More recently, Barbe *et al.* (2015) devoted a book to describing the outlines, the rules of operation and the governance of what they characterize as an "unidentified scientific object".

Since the creation of Wikipedia almost two decades ago, other self-publishing platforms have also emerged, offering amateurs new writing formats in a perspective of openness and sharing of content. But just as all amateur practices, in the field of writing as elsewhere, they do not participate in a so-called free logic; the operating methods of these platforms that give them life are far from being homogeneous and do not necessarily work towards building a common goal. Some, rightly, like Serge Proulx, are concerned about the capacity of contributory practices to surviving and developing in an ecosystem dominated by companies controlling access to content and guided solely by a commercial logic: "The contribution economy needs the expression of the subjectivities of multitudes. But, at the same time, in this expressive force of the multitudes, is contained a potential subversion of Order" (Proulx 2014, p. 29, author's translation).

This leads us to ask the following questions, which will justify the opening of future reflection: Can all these platforms claim such a commons status to the written word? What is their place in the digital book industry? What are the outlines of this new economy of cultural commons that is unfolding in the digital ecosystem of the written word?

## 4.2. Wattpad: a common narrative of the misguided written word

The arrival of social writing platforms on the Web is one of the novelties in the book ecosystem. The social site Wattpad, created by Allan Lau and Ivan Yuen in Canada in 2006, pioneered this new writing format. It specifically targets young people (more than 80% of users are under 25 years of age) by inviting them to write stories and share them with a community of readers (who can also be contributors). Most of the contributions are in the form of stories in the form of serial novels with short chapters (2,000 words on average). The reader has the opportunity to comment on the stories as they are being written, which appear legibly on the right-hand side of the main text. The founders of this platform were determined from the outset to make their activity accessible on cell phones. Their goal was to create an international community that would transform the reading and writing experience. In 2019, the figures given by the platform showed a community of 70 million users, 4 million authors and 20 million stories written in 30 different languages.

Allan Lau has a typical profile of a web entrepreneur. With an engineering background and after a few years spent at Symantec as Senior Development Manager, he co-founded his first company, Tira Wireless (2001–2007), specializing in the development of mobile publishing applications and financed by various investors for an amount of $31.5 million. In addition to his role as CEO at Wattpad, he is involved in various web start-up projects as a mentor and as a founder of investment funds. This short biographical account has its importance in the evolution of the platform, as we will show.

### 4.2.1. *The use of CC licenses: a hidden reality*

From the outset, the platform offers contributors the opportunity to publish their story under a Creative Commons (CC) license, offering all possible options without restriction. However, if we look at the page dedicated to licenses, we can see that it also offers the possibility, for the contributor, to choose a standard copyright all rights reserved license. This license is, moreover, placed in the drop-down menu before all other licenses. A detailed analysis of the site does not show any commitment of the platform for open licenses.

We can therefore hypothesize that it contributes indirectly to free culture by offering the possibility for contributors to choose a CC license, but this cannot be said to reflect a militant commitment in its favor that leads to the creation of a written commons. The use of such licenses is not an end in itself for Wattpad. Rather, it is an opportunistic choice that encourages reader engagement and story sharing and is an effective lever to increase the size of their community. In addition, it can also be seen that the search engine does not allow a targeted search to select only stories under open CC-type license.

In 2014, Wattpad and the Creative Commons association announced that the platform's community had become the world's largest provider of

CC BY 4.0-licensed text available to creators and remixers around the world: "July 21, 2014 – Wattpad, the world's largest community of readers and writers, has upgraded to Creative Commons (CC) 4.0 licensing options to give creators around the world the ability to search millions of stories to remix and reimagine. It is the largest implementation of CC 4.0 by a social media platform"[9]. This text published on the association's website actually aims to promote this type of license in the field of the written word, which, let us not forget, is the pillar of the deployment of free culture in the digital ecosystem. A figure is given to illustrate this press release: it is announced that more than 300,000 stories are shared under a CC license. Wattpad appears at the top of the ranking of digital platforms using such licenses. At first glance, this may seem like a huge number, but when you factor it into the platform's existing 20 million stories, it represents less than 2% of the stories published. To complement these figures, it would have been interesting to have longitudinal data on CC-licensed stories from 2006 to the present that would reveal the true relative influence of open versus closed licenses. In any case, Wattpad does not follow the trend of some other cultural platforms that only offer open licenses (such as the music platform Jamendo).

## 4.2.2. A progressive attraction to the attention economy

In order to continue our exploratory study on the eligibility of Wattpad to the status of written commons, we must now turn to the economic model of this platform. Originally and during the first years of its existence until 2013, it was part of a non-market ecosystem. The founders did not seek to monetize the site or to offer remuneration to history contributors. However, we can observe a change in direction, gradually moving the platform towards the commercial economy. In 2013, it offered readers the opportunity to contribute to the publication of the most popular stories through fanfunding[10] campaigns. This was made easier by the fact that each story had three popularity measurement indicators associated with it: the number of views, the number of votes and the number of comments. It was also possible to obtain rankings by type of story. Finally, once the story was finished, it was made available to readers free of charge for sharing. This was at the heart of an attention economy that marked the beginning of monetization for the

---

9 See https://creativecommons.org/wp-content/uploads/2014/07/wattpad_cc4.pdf.

10 See www.cnetfrance.fr/news/wattpad-innove-avec-le-fan-founding-le-financement-de-projets-d-ecriture-par-les-lecteurs-39793171.htm.

platform, taking 5% commission on the sums obtained by the authors when their projects succeeded. However, this was a monetization path that was still part of the non-market economy in the sense that, through this economic model, the objective remained for the platform to cover its fixed costs in terms of hosting and operation and not to make a profit.

This first monetization solution lasted only for a limited time, but the platform then moved on to another path, this time with a clear business focus, as the founders pointed out: "Now that our community and engagement numbers have grown exponentially ... we've begun to explore new paths to monetization, including branded content/native advertising and licensing."[11] Beginning in 2016, the platform began to give contributors the opportunity to introduce native advertising in their stories as an indirect way to remunerate them. Interested authors can thus supplement their own revenues by including advertisements in their story chapters. Every advert seen by readers brings in a little money: "Readers have always encouraged their favorite authors with messages, comments and votes. Now they can support them in a way that increases their income, without having to pay money out of their own pockets", Allan Lau said in an interview[12]. For users who do not want advertising, Wattpad later introduced a premium service with a $5.99/month[13] subscription in 2017. This also corresponds to a turning point in the company's development, which obtained financing of 70 million dollars after three fundraising campaigns with private investors to support its goal of reaching 1 billion users. A total of 145 people are employed in the company.

The choice to introduce advertising as a funding model is not neutral. It may jeopardize one of the pillars of free culture, as it may create bias in the motivations of contributors who are now more attracted by the search for profit than by voluntary contributions in a spirit of reciprocity, as the platform offers them a space for visibility and sharing. This reveals the line of tension between the hybrid economy and the commercial economy, as we have already pointed out in Part 1. If the rules of governance change, then this

---

11 See https://venturebeat.com/2014/04/08/social-reading-platform-wattpad-announces-massive-48m-financing-round/.

12 See www.actualitte.com/article/lecture-numerique/remunerer-les-auteurs-wattpad-avec-des-pub licites-integrees-a-leurs-histoires/66491.

13 See www.actualitte.com/article/lecture-numerique/wattpad-met-en-ligne-une-version-payante/ 85473.

may also induce a change in individual incentives to produce works on this type of platform. As Michel Bauwens pointed out about the YouTube platform, "Initially, people put films on this platform to show their video to the world because they are looking for recognition, wanting to improve their reputation, become famous… Whatever motivates them. It doesn't matter what motivates them. But as soon as you start paying these people for these specific little movies, they're going to start shooting them to make money. A capitalist mentality and logic in contradiction with the peer-to-peer mentality" (Bauwens 2015, p. 55, author's translation).

### 4.2.3. Strengthened cohabitation with publishers: the announced end of free culture

In economic terms, Wattpad's latest shift away from the spirit of free culture is taking it even further away from this said spirit. Quite quickly, it has become a pool of potentially publishable authors for traditional publishers. Some publishing houses, in France and elsewhere, have quickly become interested in this platform by keeping a daily watch on it. Isabelle Vitorino, publisher for Hachette Romans, clearly states this position:

> Being interested in this content quickly proved to be essential to remain in tune with our readership. When such a phenomenon emerges, every publisher wonders "Why didn't I think of it before?" So we rolled up our sleeves to dive into the sum of publications on Wattpad, which is huge! Fortunately, the quality of the texts we found there is well worth the time invested. The site is teeming with new talents[14].

In 2014, one of Wattpad's contributors, Anna Todd, was a major success following the continuous writing of a fanfiction, *After*, under the pseudonym @imaginator1D, with the singer of the group One Direction as the main character. Encouraged by the growing community that followed the story of this soap opera, several publishing houses in search of new talent sought to contact her to turn it into a print publication and submit a publishing contract. Simon and Schuster eventually obtained the print publishing rights in the United States; the publisher Hugo and Cie obtained the rights in France and Éditions de l'Homme in Quebec. Subsequently, the rights were

---

14 See www.myboox.fr/edito/article/tendances/wattpad-les-auteurs-de-la-plate-forme-numerique-lassaut-du-livre-papier.

bought back by Paramount for a film adaptation. To this day, the chapters remain available online. Overall, 17 stories are published on her account, which has more than 1.7 million subscribers. Most of them are copyrighted. For the others, there is no precise indication, which is quite possible knowing that an author has the possibility in the "copyright" tab to check the "not specified" box. But, in this case, the copyright prevails. It is also interesting to note that this author is at the heart of a community which, in turn, has reused her novels by giving rise to other stories (e.g. in Snapchat format). All of these stories can be considered as forms of remixing in the world of fanfiction. From a legal point of view, the contributors should have asked Anna Todd for permission before reusing the content of her novels. We can assume that this is a tacit agreement of Anna Todd, who would have the possibility to sue them in court as we mentioned in the section on the obstacles to the deployment of amateur fanfiction and thus free culture on the Internet as developed by Lessig.

Finally, we can also point out that while the platform is increasingly presented as a springboard for obtaining a contract with a publishing house, there is a risk of diverting the initial spirit. The audience, which is the basis of the attention economy's market value, is gradually seen at the heart of the platform's economic model. Various elements will, moreover, confirm that the link between the hybrid economy and the traditional commercial economy, which was considered by Lessig and Benkler as one of the levers for the development of free culture in the digital ecosystem, is a very unstable balance. In 2016, Wattpad Studios was created, a subsidiary responsible for adapting the most popular stories to the entertainment market. The goal was to connect the most popular authors with companies in the entertainment industry, such as NBCUniversal. In this new type of partnership, authors are remunerated via a classic publishing contract and the platform on the basis of their activity as co-producers. Wattpad plays a new role as an agent by helping their authors find publishers and producers: "Last year alone, the books of three of our authors made the *New York Times* bestseller list. I should point out that we do not retain any rights to any of the content that is published on our platform. The author retains all rights."[15]

In 2019, a new step was taken. Wattpad created its own publishing house with an investment in machine learning to identify the most promising trends and also to experiment with new formats such as virtual reality or games. The

---

15 See www.journaldunet.com/media/publishers/1194433-allen-lau-wattpad/.

goal is now to become the factory of the entertainment industry. Wattpad has also expanded worldwide an experimental program called Wattpad Paid Stories, which allows user-readers to directly reward authors by buying chapters of their stories. A selection of around 100 stories is made by the platform upstream. This strategy of remunerating authors directly by readers ultimately amounts to falling back into a classic editorial model, with these stories only being available once they are purchased.

The evolution of its strategy and, in particular, of its economic model over the last decade attest that Wattpad has gradually left the hybrid economy, the archetype of free culture, to join the ranks of the traditional commercial economy. Several lessons can be drawn from this case study. First, it is not enough for a platform to allow the use of CC licenses for it to be considered a cultural commons. Similarly, free access to content is not a convincing indicator either, since such access does not guarantee the right to copy and reuse. Finally, the choice that has been made to try to coexist and then become a key player in the publishing industry has finally distanced this platform from the promotion of free culture. For all these reasons, we have chosen to qualify Wattpad as a misguided commons.

## 4.3. Self-publishing and free culture: a multifaceted face

The self-publishing sector has literally exploded with the progressive arrival of many players who have created platforms that offer the author the possibility to publish his or her book easily and simultaneously find an audience. Among these actors, we have selected three that, because of the possibility they offer to authors to publish their work under a Creative Commons license, deserve special attention.

### 4.3.1. *The Lulu platform: open source for the book market?*

We will begin with the Lulu[16] self-publishing platform. Several clues let us imagine that this is a platform that is a good place for the deployment of amateur free culture. First of all, it was created in 2002 by the co-founder and former CEO of the free software company Red Hat, Bob Young. In an interview given on the website of the Creative Commons association, his manager claims a strong kinship with the approach of free software aimed at

---

16 See http://lulu.com.

creating what he calls the open source book market: "The Internet liberalizes the book publishing business as free software has done for computer publishing: it is the open source book market."[17] The platform makes it possible to liberalize the publishing market by offering the possibility for creators to keep their rights, leading to a loss of absolute control over publishers. On the other hand, Young was one of the co-founders[18] of the Center for Public Domain at Duke University, a non-profit association that was created to support important negotiations on intellectual property, patents and copyright law, and the management of the public domain for the common good. It was mentioned earlier how some members of the Berkman Center, like James Boyle, have played a fundamental role in this association, which is at the heart of American activism for free culture. We might be led to believe that we will find on this platform creation under the CC license, in free access.

But the reality is more nuanced. Indeed, it is possible for an author to be able to publish his creation under this open license. But if we go to the platform, the number of digital books published with such a license remains marginal and we would even say that it does nothing to make these creations easily accessible on the readers' side. From the first page, the timeline presents four tabs, "buy", "create", "sell", "learn", which leave no doubt about the commercial nature of the platform. The search engine does not allow a specific sorting by license type either. Finally, the works protected by copyright (the vast majority) are not even accessible for reading. There is clearly a discrepancy between their apparent will to be part of the free software filiation and their strategy. An interview conducted by the Creative Commons association in 2006 reveals this contradiction. Faced with the fact that about 300 creations are under CC license (which is low compared to the total number of existing ones), their answer reveals an opportunistic vision linked to the use of this type of license:

> Demand from the creator community is the reason Lulu offers those licenses! Despite being early supporters of Creative Commons, we were slow to offer the licenses on our site because our team was so busy with other features. But eventually we had to make Creative Commons options available, because as a company we pay close attention to what members of the Lulu community talk about and request. While

---

17 See https://creativecommons.org/2006/05/17/lulu/.

18 This is mentioned in his Wikipedia page.

the flexibly licensed works constitute a minority of the total number of books published on Lulu.com, the folks who use them carry a lot of weight with us[19].

The heart of their project is elsewhere.

In another interview, Young explains the purpose of the self-publishing project: "Most manuscripts are rejected by publishers, not because they are bad, but because they would reach too small an audience to become profitable books". He adds, "When a manuscript is accepted, the author is often disappointed by the very low income it will generate"[20]. The innovative aspect of their project is that they were the first to adapt the Long Tail model to the world of books, referring to people who have something to say but whose audience is too small to have traditional publishers as the main target. Each Internet user can publish his or her book, which he or she sends to the platform. The author, who remains the owner of the rights related to the content of the book, does not have to pay anything. He or she sets the selling price of the book, which can be ordered on the Web. At each sale, Lulu deducts the manufacturing cost (about 2 euro cents per page) and then pays 80% of the remaining amount to the author. For a 200-page book sold for 20 euros, the author will receive about 12 euros per copy purchased, compared to less than 2 euros from a traditional publisher. Richard Brown, one of the best-known authors on Lulu, who sells only 1,000 copies, collects $32,000 from his annual sales.

Lulu's objective is therefore not to compete with publishers, but to create a new market segment that is financially very promising for them. It is not a question of ensuring that real best-sellers emerge via this model: "If a million people publish a book that sells 100 copies each, that can make Lulu the world's leading publisher"[21], says founder Bob Young. Very quickly, Lulu achieved a significant turnover (several million dollars) and is now claiming the online publishing of more than a million books, half of which are fiction. The site receives an average of 100,000 visits per day and has a community

---

19 See https://creativecommons.org/2006/05/17/lulu/.

20 See www.zdnet.fr/actualites/lulucom-le-premier-editeur-en-ligne-ouvert-a-tous-les-manuscrits-39360859.htm.

21 See www.zdnet.fr/actualites/lulucom-le-premier-editeur-en-ligne-ouvert-a-tous-les-manuscrits-39360859.htm.

of 1.32 million members. There is no doubt in light of this exploratory analysis that this platform, which claims to be the open source of books, does not contribute in any way to the development of free culture and of the written commons in the digital ecosystem.

### 4.3.2. *In Libro Veritas and Framabook: free book editions*

Around the same time, in 2006, the In Libro Veritas[22] (ILV) website was created in France by Mathieu Pasquani. Like the founder of Lulu, he was inspired by the spirit of free software and applying it to the world of culture and books in particular. In an interview on the Onirik blog, he states, "Quite naturally the idea of associating these works with free licenses, which were still in their infancy, imposed itself on me just as free software had imposed itself on coders for 20 years. Thanks to these free distribution licenses (Free Art License, Creative Commons, GFDL, GNU/GPL) the certainty of being able to distribute knowledge while legally securing the rights and duties of authors and readers was finally within our reach"[23].

This site presents itself as a new kind of publishing house and the first to offer free books. It allows everyone to write and publish works online and the selection is made *a posteriori* by the community of readers who "select" the works most worthy of being read. This costs nothing to the author or the reader, who can access it free of charge and download practically all the texts (a minority being under copyright). The author can find remuneration via the paper sales of his or her digital book and receives 100% of the profits. He or she can also see visibility increased by choosing a free license, which attests to the platform's commitment to the development of free culture. As for the platform, it is remunerated on additional services by offering support in the highlighting of texts on preferred sites or by offering print-on-demand services or advanced author account administration. In terms of notoriety, it claims 5,350 authors, 29,359 works, 1,617 subscribers and a total number of downloads of 2.3 million. These figures show that in its 15 years of existence, this platform has experienced relative growth, but that still remains very confidential at this stage. Moreover, to date, ILV seems to be a relatively dormant site: the latest publications over the past year

---

22 See www.inlibroveritas.net/le-concept.html.

23 See www.onirik.net/Interview-de-Mathieu-Pasquini.

are 11 in number, the vast majority with very short formats, limited to about 10 pages. The associated social networks have not been functioning since 2015.

ILV seems to have been challenged by the opening of a platform called Tramenta, with a rather similar foundation, created in 2011 by ILV's co-founder, Thomas Boitel, who is in charge of its development. This site contains more than 5,000 works that can be read for free or downloaded as e-books, in PDF, epub or Kindle-compatible formats (including a number of works in the public domain). This platform differs from ILV in that it offers Internet users the possibility of publishing their text under copyright. In addition, it has chosen to broaden its scope of distribution by offering a support service for major online bookstores (Amazon, Apple, Google, Fnac, etc.).

In both cases, we can see the existence of an authentic literary community composed of Internet users who, for the most part, have themselves published books on these platforms. A quick glance at the nature of the comments reveals real cross-exchanges that go far beyond the simple mention of "I like this" and a closeness between Internet users forged through their mutual reading. For all these reasons, these platforms illustrate the archetypal written commons, finding a place in the digital book ecosystem. They undeniably promote freedom of expression and sharing around published texts. Finally, their economic model avoids any deviant form in the behavior of Internet contributors, as can be the case when the platform chooses to be at the heart of an attention economy.

This work is only exploratory at this stage and should, however, be extended by a more detailed analysis that would make it possible to evaluate, for example, the distribution of the texts according to the type of license chosen, the degree of commitment of the contributors as commoners (their profile can vary from simple "consumer" to that of contributor committed to free culture) and finally the modalities of the functioning of the governance.

Before closing this chapter, we would like to mention one last platform which, in our opinion, is the one most committed to the development of free culture and the construction of a written commons. This is Framabook, which is a collection of free books published by the Framasoft[24] association. It is based on a collaborative work method between the author and the association's volunteers, the latter having a reading committee and an editorial committee. In terms of genre, it is rather specialized on manuals (often computer science), essays and comics. As it explicitly underlines on its site, the choice was made to propose to its contributors only licenses "that are part of the free culture and participation in the common goods", or licenses that ensure the user a free use, copy, modification or redistribution of the work and its derivatives. However, it should be noted that two types of CC licenses are excluded, those with an NC ("non-commercial") clause and those with an ND ("no modification") clause. This willingness to promote only very permissive licenses (Free Art License 1.3, CC BY SA 3.0, CC BY 3.0, or even CC 0) is explained by the desire to promote an ecosystem of sharing for the written word. Stemming from the popular education movement, the platform's objective is to enable dissemination to the greatest number of people. However, according to the association, the NC clause contravenes this principle. This choice can be questioned and the exclusion of certain CC licenses does not in itself constitute a stronger commitment to free culture. It is also specified that the authors themselves participate in the modalities of the editorial policy during general meetings.

In terms of the economic model, free does not automatically mean completely free. Framabook proposes an economic model based on the primacy of distribution and fair remuneration of authors. It proposes a remuneration of the authors via a system of donations and the possibility of selling digital books in paper format. In this case, the buyer is sent back to the Lulu platform to make the purchase. The statistics available[25] online reveal a relatively small stock of published texts, since there are only "42 books or other documents available" for a relatively large number of downloads (3.3 million since the opening of the site), but with a computer book alone having been downloaded 2.1 million times. We can also note the absence of exchanges in the form of comments or assessments on the site, unlike the two

24 Coming from the educational world and now focused on popular education, the Framasoft association is above all a network of projects, the first of which, the Framalibre directory, dates back to 2001. These projects are led by people working together around the same desire: to promote digital freedoms. The respect of the fundamental freedoms of users, guaranteed by legal contracts (free licenses), is at the heart of the library movement and ensures that humans remain in control of the digital tool.

25 See https://framastats.org/#Framabook.

other platforms mentioned above. However, this does not mean that there is no active community around this association, but it is, in our opinion, upstream, at the level of the publishing association Framasoft.

Most of the association's income is made up of donations. Income from the sale of books provides working capital to purchase stocks of printed matter. Finally, one last point is worth mentioning as it relates to one of the issues discussed in Part 1, concerning the remuneration of volunteer commoners. Indeed, the Framabook site highlights the introduction of what Framabook calls "enhanced volunteering", which makes it possible to quantify volunteer activity by taking into account the "non-financial" flows of Framasoft project activity. According to the association, this makes it possible to justify the fact that, while generating income, it remains a predominantly non-profit organization. Thus, any volunteer who has participated in translation, web development, conference, proofreading and system administration tasks is encouraged to evaluate them based on qualitative elements (type of task or project) and quantitative elements (number of hours spent and estimated salary corresponding to the tasks performed).

This survey, carried out among these different self-publishing platforms, which seemed to us to be eligible for the status of written commons because of the possibility they offer users of using Creative Commons licenses, was rich in lessons.

First of all, the possibility given to its community to use a Creative Commons license does not appear to be a sufficient condition to conclude that the platform promotes free cultural creative practices and jointly the emergence of the written commons. The Wattpad platform, by being the largest provider of CC licenses in the field of writing, has often been valued for its contribution to the development of free culture. It contributes to this quantitatively, but, we could say, in spite of it or at least without having had the deliberate intention to do so. This platform makes a diverted and opportunistic use of it. Its strategy of offering creators both exclusive licenses (such as copyright) and open licenses (such as CC) aims above all to obtain a large user base in order to hopefully trigger increasing returns on adoption. The evolution of Wattpad's economic model, with an increasing monetization of the content offered by its community, reinforces this hypothesis. The same is true for the Lulu self-publishing platform, which has clearly forgotten its declared affiliation with the free world and has preferred to consolidate a niche position in the book industry by also playing on the positive externalities brought about by a growing user base. The search for profit has gradually moved these platforms away from what Berkman Center jurists call the hybrid cultural economy to definitively orient them towards a traditional commercial economy.

Beyond the use of open licenses, what in our view is the fundamental issue at stake for a written commons is the way the platform promotes free creative practices while avoiding dissolving into the attention logic dominating the digital ecosystem. With Wattpad, we were able to see how porous the boundary between the hybrid model and the commercial model is. Moreover, as soon as we put contributors in a situation where they are given the opportunity to make gains from their amateur practices, or if they are given hope of becoming the next Anna Todd, this balance threatens to collapse. The boundary between giving and calculating appears more indeterminate and risks leading to a breach of the implicit social contract.

Conversely, the other platforms studied, ILV and Framabook, display a strong commitment to free culture and offer their users only open CC-type licenses. This choice is part of their DNA, one of the founding values of their project. One has chosen to remain in the non-market world (paying itself only through donations); the other has chosen the market universe by offering additional paid services but in a hybrid logic, in the sense that it does not make the search for profit its primary motivation. However, we have seen how difficult it was for both of them to develop, or even to sustain their activity. At this stage, the question of cohabitation with the traditional actors of publishing does not even arise. Their activity remains very confidential. It neither threatens nor seems to arouse any interest whatsoever on the part of the actors of the book industry. On the other hand, these platforms have been able to create an authentic participatory community that nourishes the creative practices of their users. The latter do not come to these platforms by chance. It is a deliberate choice that reinforces these platforms to be authentic forms of the written commons in the digital ecosystem.

# Conclusion

Without an initial or over-determining position, we have explained the contemporary intellectual movement of the commons in the digital cultural ecosystem by identifying the places of production, the spaces of reflection and the lines of tension that characterize it. Our intellectual journey was guided by the search for enlightening answers to the following questions: What is a cultural commons? What is its identity? What are the characteristics of a cultural commons economy?

In the digital ecosystem, cultural commons are not easy to grasp. They are rarely named as such by their own "designers". While they can be identified in the form of digital platforms, information infrastructures that share cultural resources produced by contributors, they are more than that. For they also refer to the institutional arrangements that promote the sharing and enrichment of these cultural resources (artistic, scientific, heritage, etc.). These institutional arrangements that address the conditions for sharing them and the nature of the community and associated governance are of paramount importance, as they allow cultural commons platforms to stand out in an essential way from the many so-called sharing or collaborative platforms that populate the digital ecosystem.

Cultural commons exist only through the action of contributing users who choose to share on a digital platform the product of their artistic and intellectual creation, or to contribute to the creation of a collective work shared on a voluntary basis. Their status does not matter. In the artistic field, these platforms bring together the creative contributions of amateurs, but not only this. Conversely, on scientific commons platforms, we find mostly professionals, but not exclusively. These contributors, also known as commoners, agree to make their intellectual creations available to others in the form of sharing without financial compensation. The voluntary nature of their contributions, whether individual or collective, is an essential property of the cultural commons. This is as much the case for Wikipedia contributors as it is for the social writing platform Wattpad or the Framabook free book publishing platform. The same is true for the authors of scientific articles deposited on open archive platforms. In the field of heritage, digitized works in the public domain, the question does not arise in these terms.

These cultural resource platforms are eligible for a commons status because they propose to share them through the establishment of formal rules in the form of a set of rights concerning the conditions of access, appropriation and reuse of these resources. These institutional arrangements are not the same from one platform to another and are closely linked to the nature of cultural resources and their conditions of production, circulation and consumption in the digital ecosystem. For creative commons in the artistic field, this translates into the provision of open licenses of the Creative Commons type for the contributor. These are multiple and each one refers to more or less permissive conditions of sharing, appropriation and reuse. For the scientific commons, the conditions of access cover a very broad spectrum, from a simple reading of the texts to very permissive conditions of reuse that encourage the exploration and dissemination of scientific knowledge. For platforms offering digitized heritage resources in the public domain, there are several possible legal arrangements, even if dedicated open licenses have been created following the example of the Etalab license to promote their reuse.

Platforms for cultural commons are characterized by the existence of a community with a diffuse outline, with access to cultural resources not being reserved for a specific category of commoners. It brings together not only the contributors, but also all potential users who appropriate their content, these two categories not being themselves exclusive of each other. This community associated with a cultural commons is defined by access rights that are assigned to it through open licenses protecting the available resources. In the cases studied, all users have the same rights, regardless of their status and degree of commitment to the effective contribution, leading to the production of the commons. However, we do not believe that this characteristic is intrinsic to the definition of a cultural commons. We can imagine platforms that choose to modulate the conditions of sharing according to the nature of the users.

The third essential dimension of cultural commons is their governance. At the organizational level, it can be centralized around the platform owners, which is often the case for creative commons in the artistic field. We can also find cooperative organizational forms, such as Wikipedia, which is often cited as an example of self-managed governance with community members. However, this is not a norm that has been imposed on all cultural commons. In the field of heritage commons, in some of the cases studied, governance is based on a hybrid organizational mode materialized by a partnership between the cultural institution and the private actor participating in the co-production of digitized heritage. The scientific commons also often rely on hybrid forms of governance articulating different entities (universities, researchers, etc.) that discuss and elaborate the rules of use collectively, as well as on more centralized forms with a single regulatory entity. The organization of governance is therefore strongly linked to the nature of the cultural resource and the nature of its ecosystem. In the case of heritage commons, the cultural institution must not only ensure the development of the platform; it must also bear the cost of creating the resource units, and manage the implementation of the digital cultural heritage production process. In all other cases, these resource units are digitally native and are contributed voluntarily by the community.

At the institutional level, governance is defined by a set of collective rules aimed at regulating the conditions of access and use for commoners and, more broadly, for the community as a whole. Here, unlike land commons, the social dilemma is not a risk of overexploitation of the resource resulting from the actions of stowaways or non-cooperation on the part of certain members of the community. On the contrary, since knowledge, whatever its content, is a non-rival resource, and is itself enriched by existing knowledge in a dynamic perspective, there is no risk of overexploitation. The objective of the governance of a cultural commons platform is to foster continuous dynamics of creation of such resources. Fostering a climate that encourages commoners to contribute regularly to the cultural commons is an essential dimension of this. In the case of artistic creative commons, the importance of respecting an implicit social pact for governance based on a principle of reciprocity or mutual benefit between their representatives and the platform's contributors was stressed, so that the latter are always inclined to contribute. For common heritage, as there are no contributors strictly speaking, this role being devolved to the cultural institutions and partners in charge of the process of digitization and the platformization of heritage, it is directly incumbent on them to implement the conditions for continuous enrichment. The importance of editorialization and digital mediation strategies in this area has been shown, which encourage users to play this role through actions of reuse and enrichment of heritage content. For the scientific communities, it appeared that some major obstacles hindering the individual incentive to deposit articles in open archives are due to exogenous variables, such as research evaluation methods.

Faced with the enclosures voluntarily created by various economic actors on knowledge (scientific, artistic, heritage) with a view to the emergence of new markets or the protection of existing ones, the development of cultural commons platforms demonstrates that plausible alternatives are possible and are likely to pave the way to another knowledge economy. This cultural commons economy is characterized by three founding principles: shared ownership of cultural resources, social value generated by voluntary contributors and governance oriented towards the continuous enrichment of cultural resources.

The modalities of cohabitation with the traditional economic actors of the cultural industries and, in particular, with publishers and existing distribution platforms constitute a central challenge for cultural commons platforms. Indeed, even if a large number of them are part of the non-market economy and therefore not part of the cultural sector, they *a priori* do not pose a direct threat to the economic players in place; they are all intended to become a sustainable part of the cultural digital ecosystem.

In the artistic field, creative commons platforms (written or otherwise) are not direct competitors of commercial cultural platforms. From the outset, they have been aimed at amateur creators and not at professional artists, offering them unprecedented opportunities to showcase and share their creations. But it has been clear that these two worlds have not always remained foreign to each other. Wikipedia has completely destabilized the encyclopedia market without any intention of doing so. In the field of writing, while self-publishing and social writing platforms, such as Wattpad, did not arouse any particular interest among publishers, at least initially, gradually, and with the growing popularity of some of them, some have moved closer to them to tap into the pool of amateur writing in order to identify new talent. In the case of Wattpad, this had an impact on the platform itself, which gradually sought in return to become a new player in the commercial cultural economy, jeopardizing the implicit social pact with its community of voluntary contributors. In the scientific field, the first open archives were developed by researchers to improve their scientific conversation through a new and more efficient digital ecosystem for sharing their pre-print articles. But at the same time, all these experiments very quickly became part of the open access movement aimed at imposing a new order of scientific communication that would put an end to the growing reign of scientific publishers. It is certain that their very existence has contributed to the change of the entire ecosystem, which in turn has required publishers to adapt to this new environment of scientific communication and these new modes of intermediation. To date, this environment is far from being stabilized, and major changes are to be expected due to the intervention of European governments, particularly in favor of massive support for this OA movement.

In the field of heritage commons, the question of cohabitation with economic actors is played out at another level, because here the cultural commons is not a native digital resource. We have been able to show that hybrid private–public co-production methods are not intrinsically harmful to the enrichment of the commons, even if they entail risks of enclosures that do not always come from private actors as we would tend to believe.

The question of cohabitation between the actors of cultural commons and those of the commercial economy has also revealed the importance of the exploitation conditions of the social value created on these commons platforms, which rests, let us recall, on the voluntary contributions of the community. This created social value poses two types of challenges: the methods of remunerating volunteers as a condition of enrichment and sustainability of the commons and the conditions of exploitation of this social value by the owners of the platform through the chosen economic model.

Remuneration for voluntary contributions emerged as a key dimension of the issue of value extracted from cultural commons. Although there is no question of introducing a system of direct monetary remuneration, options have been considered to guarantee the sustainability of the commons. Here, however, tensions have been highlighted, as the solutions envisaged each convey, in the background, a singular and hardly reconcilable vision of the political economy of the cultural commons and its future within a rapidly growing platform capitalism governed by attentional logics.

The question of the choice of economic model also emerged as a central issue. A certain number of cultural commons platforms are part of the non-market economy with funding based on donations, sponsorship and/or public money. This is not the case for all of them. In this second case, the cultural commons economy is a hybrid economy where the choice of the appropriate economic model is a key issue. In particular, opting for a potentially highly remunerative economic model that takes advantage of the size of the community, such as the advertising model, can be very risky because it can lead to a breach of the reciprocity pact between the owners of the platform and the voluntary contributors. All these issues have emerged at the heart of the challenges for the creative commons economy in the artistic field. For the heritage commons, this question is all the more sensitive since the question of prior financing of the digitization of heritage in a context of rationalization of cultural spending arises. This double constraint has paved the way to creative forms articulating public/private logics. In the scientific field as in the heritage field, the actors who are at the origin of these commons platforms do not have the vocation to benefit from them. However, in the near future, they will have to take up the challenge of proposing an offer of information and communication services that is sufficiently attractive to attract and retain contributors and users. In each case, a balance needs to be struck based on complex hybrid arrangements. Different options are possible to promote the enrichment of the common good. Some of them are most likely still to be invented.

While it is undeniable that many areas of uncertainty and tension remain to this day on the foundations and future of the cultural commons economy in the digital ecosystem, we hope that the reflection in this book will inspire others to pursue this line of research.

# References

Aigrain, P. (2005). *Cause commune, l'information entre bien commun et propriété*. Fayard, Paris.

Aigrain, P. (2012). *Sharing Economy, Culture and the Economy in the Internet Age*. Amsterdam University Press, Amsterdam.

Alchian, A. and Demsetz, H. (1973). The property rights paradigm. *Journal of Economic History*, 33, 13–27.

Alix, N., Bancel, J.L., Coriat, B., Sultan, F. (2016). *Vers une république des biens communs ?* Les liens qui libèrent, Paris.

Arrow, K.J. (1962). Economic welfare and the allocation of resources for invention. In *The Rate and Direction of Inventive Activity: Economic and Social Factors*, National Bureau of Economic Research (ed.). Princeton University Press, Princeton, New Jersey.

Barbe, L., Merzeau, L., Schafer, V. (2015). *Wikipédia, objet scientifique non identifié*. PUF, Paris.

Barberousse, A., Laloë, F., Guyon, E. (2003). Les immenses enjeux de la communication scientifique directe. *Genesis (Manuscrits–Recherche–Invention)*, 20, 177–183.

Bauwens, M. (2015). *Sauver le monde. Vers une économie post-capitaliste avec le peer-to-peer*. Les liens qui libèrent, Paris.

Bauwens, M. and Kostakis, V. (2017). *Manifeste pour une véritable économie collaborative. Vers une société des communs*. Éditions Charles Léopold Mayer, Paris.

Bauwens, M. and Niaros, V. (2019). *Value in the Commons Economy: Developments in Open and Contributory Value Accounting*. P2P Foundation, Amsterdam.

Bellon, A. (2017). Le hacker et le professeur. Mise en débat de la propriété intellectuelle sur Internet aux États-Unis. *Raisons politiques*, 3(167), 165–183.

Benabou, V.L. (2014). Rapport de la mission du CSPLA sur les "œuvres transformatives". Report, Commande du ministère de la Culture et de la Communication au CSPLA, Paris.

Benhamou, F. (2014). *Le livre à l'heure numérique. Papiers, écrans, vers un nouveau vagabondage*. Le Seuil, Paris.

Benhamou, F. and Farchy, J. (2014). *Droit d'auteur et copyright*, 3rd edition. La Découverte, Paris.

Benkler, Y. (2002). Coase's Penguin, or Linux and "The Nature of the Firm". *The Yale Law Journal*, 112(3).

Benkler, Y. (2003). The political economy of commons. *European Journal for the Informatics Professional*, 4(3), 6–9.

Benkler, Y. (2006). *The Wealth of Networks*. Yale University Press, New Haven.

Benkler, Y. (2011). *The Penguin and the Leviathan. The Triumph of Cooperation over Self-interest*. Crown Business, New York.

Benkler, Y. and Nissenbaum, H. (2006). Commons-based peer production and virtue. *The Journal of Political Philosophy*, 14(4), 394–419.

Bollier, D. (2014). *La renaissance des communs. Pour une société de coopération et de partage*. Éditions Charles Leopold Mayer, Paris.

Bomsel, O. (2007). *L'économie du gratuit. Du déploiement de l'économie numérique*. Folio, Paris.

Boyle, J. (2003a). The second enclosure movement and the construction of the public domain. *Law and Contemporary Problems*, 66(33), 33–74.

Boyle, J. (2003b). Foreword: The opposite of property. *Yale Journal of Law*, 9(66), 1–2.

Boyle, J. (2008). *The Public Domain, Enclosing the Commons of the Mind*. Yale University Press, London.

Broca, S. (2013). *Utopie du logiciel libre. Du bricolage informatique à la réinvention sociale*. Les liens qui libèrent, Paris.

Cardon, D. (2010). *La démocratie internet*. Le Seuil, Paris.

Cardon, D. and Levrel, J. (2009). La vigilance participative. Une interprétation de la gouvernance de Wikipédia. *Réseaux*, 2(254), 51–89.

Casilli, A. (2016). Is there a global digital labor culture? Marginalization of work, global inequalities, and coloniality. *2nd Symposium of the Project for Advanced Research in Global Communication (PARGC)*, Philadelphia, April.

Chartier, R. (2008). Le livre, son passé, son avenir. *La vie des idées* [Online]. Available at: http://www.laviedesidees.fr/Le-livre-son-passe-son-avenir.html.

Citton, Y. (2014). *L'économie de l'attention. Le nouvel horizon du capitalisme ?* La Découverte, Paris.

Coriat, B. (2013). Le retour des communs. Sources et origines d'un programme de recherche. *Revue de la régulation*, 14.

Coriat, B. (2015). *Le retour des communs. La crise de l'idéologie propriétaire.* Les liens qui libèrent, Paris.

Crétois, P. (2014). La propriété repensée par l'accès. *Revue internationale de droit économique*, 28, 319–334.

Da Silva, L. (2013). Genèse et description des bibliothèques numériques. *Documentation et bibliothèques*, 59(3), 132–145.

Dardot, P. and Laval, C. (2014). *Commun : essai sur la révolution au XXI^ème siècle.* La Découverte, Paris.

Darnton, R. (2009). *L'apologie du livre. Demain, aujourd'hui, hier.* Gallimard, Paris.

De Filippi, P. and Hassan, S. (2014). Measuring value in the commons-based ecosystem: Bridging the gap between the commons and the market. In *The MoneyLab Reader*, Lovink, G. and Tkacz, N. (eds). Institute of Network Cultures, Amsterdam, University of Warwick, Coventry.

Doueihi, M. (2008). *La grande conversion numérique.* Le Seuil, Paris.

Dulong de Rosnay, M. and de Martin, J.C. (2012). *The Digital Public Domain, Foundations for an Open Culture.* Open Book Publishers, Cambridge [Online]. Available at: https://halshs.archives-ouvertes.fr/file/index/docid/726835/filename/THE_DIGITAL_PUBLIC_DOMAIN.pdf.

Ertzscheid, O. (2010). L'homme est un document comme les autres. *Hermès, La Revue*, 1(53), 33–40.

Eynaud, P. and Laurent, A. (2017). Articuler communs et économie solidaire : une question de gouvernance ? *RECMA*, 3(345), 27–41.

Fallery, B. and Rodhain, F. (2013). Gouvernance d'Internet, Gouvernance de Wikipédia : l'apport des analyses d'E. Ostrom sur l'action collective auto-organisée. *Management et Avenir*, 7(65), 169–188.

Farchy, J., Meadel, C., Sire, G. (2015). *La gratuité à quelle prix ? Circulation, échanges culturels sur Internet.* Presses universitaire des Mines, Paris.

Flichy, P. (2010). *Le sacre de l'amateur. Sociologie des passions ordinaires à l'ère du numérique.* Le Seuil, Paris.

Frosio, G. (2012). Communia and the European domain project: A politics of the public domain. In *The Digital Public Domain: Foundation for and Open Culture*, de Martin, J.C. and Dulong de Rosnay, M. (eds). Open Book Publishers, Cambridge.

Guedon, J.C. and Loute, A. (2017). L'histoire de la forme revue au prisme de l'histoire de la grande conversation scientifique. *Matérialités et actualités de la forme revue*, 12, 1–24.

Guibet Lafaye, C. (2014). La disqualification économique du commun. *Revue internationale de droit économique*, 3(28), 271–283.

Hardin, G. (1968). The tragedy of the commons. *Science*, 162(3859), 1243–1248.

Hess, C. and Ostrom, E. (2003). Ideas, artifacts and facilities: Information as a common-pool resource. *Law and Contemporary Problems*, 66, 111–145.

Hess, C. and Ostrom, E. (2007). *Understanding Knowledge as a Commons: From Theory to Practice.* MIT Press, Boston.

Hollard, G. and Sene, O. (2010). Elinor Ostrom et la gouvernance économique. *Revue d'économie politique*, 120(3), 441–452.

Jeannenet, J.-N. (2009). *Quand Google défie l'Europe. Plaidoyer pour un sursaut.* Éditions Mille et Une Nuits, Paris.

Jenkins, H. (2013). *La culture de la convergence. Des médias au transmédia.* Armand Colin, Paris.

Juanals, B. and Minel, J.L. (2016). La construction d'un espace patrimonial partagé dans le web des données ouverts. *Communication*, 34(1), 17–32.

Juanals, B. and Noyer, J.-M. (2010). *Technologies de l'information et intelligences collectives.* Hermes-Lavoisier, Paris.

Lakhani, K.R. and Wolf, R.G. (2005). Why hackers do what they do: Understanding motivation and effort in free/open source software projects. In *Perspectives on Free and Open Source Software*, Feller, J., Fitzgerald, B., Hissam, S., Lakhani, K.R. (eds). MIT Press, Cambridge.

Latournerie, A. (2001). Petite histoire des batailles du droit d'auteur. *Multitudes*, 5, 37–62.

Latrive, F. (2004). *Du bon usage de la piraterie*. Exils, Paris.

Le Crosnier, H. (2006). Économie de l'immatériel : abondance, exclusion et biens communs. *Hermès, La Revue*, 45, 51–59.

Le Crosnier, H. (2012). Elinor Ostrom. L'inventivité sociale et la logique du partage au cœur des communs. *Hermès, La Revue*, 64, 193–198.

Le Crosnier, H. (2015). *En-communs : une introduction aux communs de la connaissance*. C&F éditions, Paris.

Legendre, B. (2019). *Ce que le numérique fait aux livres*. Presses universitaire de Grenoble, Grenoble.

Lescure, P. (2013). Culture-acte 2, mission "Acte II de l'exception culturelle". Report, Commande du ministère de la Culture et de la Communication, Paris.

Lessig, L. (1999a). *Code and Other Laws of Cyberspace*. Basic Books, New York.

Lessig, L. (1999b). Reclaiming a commons. Draft 1.01. Keynote address. *Building a Digital Commons*. The Berkman Center, Cambridge.

Lessig, L. (2001). *The Future of Ideas. The Fate of the Commons in a Connected World*. Random House, New York.

Lessig, L. (2006). *CodeV2*. Basic Books, New York [Online]. Available at: http://codev2.cc/download+remix/Lessig-Codev2.pdf.

Lessig, L. (2008). *Remix: Making Art and Commerce Thrive in the Hybrid Economy*. Bloomsbury Publishing, London.

Lessig, L. (2009). *Free Culture*. Penguin Books, London.

Levy, P. (1997). *L'intelligence collective. Pour une anthropologie du cyberspace*. La Découverte, Paris.

Levy, P. (2010). L'espace sémantique IEML. Vers une réflexivité de l'intelligence collective sur le web. In *Technologies de l'information et intelligences collectives*, Juanals, B. and Noyer, J.-M. (eds). Hermes-Lavoisier, Paris.

Maurel, L. (2014). Droit d'auteur et création dans l'écosystème numérique : des conditions d'émancipation à repenser d'urgence. *Mouvements*, 3(279), 100–108.

Maurel, L. (2017). La reconnaissance du domaine commun informationnel : tirer les enseignements d'un échec législatif [Online]. Available at: https://hal.archives-ouvertes.fr/hal-01877448/document.

Méadel, C. and Sonnac, N. (2012). L'auteur au temps du numérique. *Esprit*, 102–114.

Mounier, P. (2012). Les différents types d'édition numérique. *Dazibao, Revue Agence Régionale du Livre*, Spring.

Nesson, C. (2012). Foreword. In *The Digital Public Domain. Foundations for an Open Culture*, Dulong de Rosnay, M. and de Martin, J.C. (eds). Open Book Publishers, Cambridge.

Noyer, J.-M. and Carmes, M. (2013). Le mouvement "open data" dans la grande transformation des intelligences collectives. In *Les débats du numérique*, Carmes, M. and Noyer, J.-M. (eds). Les Presses des Mines, Paris.

Olson, M. (1965). *The Logic of Collective Action: Public Goods and the Theory of Groups*. Harvard University Press, Cambridge.

Orsi, F. (2015). Revisiter la propriété pour construire les communs. In *Le retour des communs*, Coriat, B. (ed.). Les liens qui libèrent, Paris.

Ostrom, E. (1990). *Governing the Commons: The Evolution of Institutions for Collective Action*. Cambridge University Press, Cambridge.

Ostrom, E. (2009). Beyond markets and states: Polycentric governance of complex economic systems. *Nobel Lecture*, December.

Ostrom, E. (2010). *La gouvernance des biens communs. Pour une nouvelle approche des ressources naturelles*. De Boeck, Brussels.

Ostrom, E. and Schlager, E. (1992). Property rights regimes and natural resources. A conceptual analysis. Workshop in Political Theory and Policy Analysis. Indiana University, Bloomington.

Peugeot, V. (2011). *Libres savoirs : les biens communs de la connaissance – produire collectivement, partager et diffuser les connaissances au XXI^{ème} siècle*. C&F éditions, Paris.

Pirolli, F. (2015). *Le livre numérique au présent. Pratiques de lecture, de prescription et de médiation*. Éditions universitaires de Dijon, Dijon.

Proulx, G. (2014). Enjeux et paradoxes d'une économie de la contribution. In *La contribution en ligne. Pratiques participatives à l'ère du capitalisme informationnel*, Proulx, G., Garcia, J.L., Heaton, L. (eds). Presses universitaires du Québec, Quebec.

Proulx, G., Garcia, J.L., Heaton, L. (2014). *La contribution en ligne. Pratiques participatives à l'ère du capitalisme informationnel*. Presses universitaires du Québec, Quebec.

Rifkin, J. (2005). *L'âge de l'accès*. La Découverte, Paris.

Rifkin, J. (2016). *La nouvelle société du coût marginal zéro*. Éditions Babel, Paris.

Robin, C. (2015). Les livres numériques au centre d'une économie de l'attention ? In *Le livre numérique au présent*, Pirolli, F. (ed.). Éditions universitaires de Dijon, Dijon.

Rochfeld, J. (2014). La propriété s'oppose-t-elle aux communs ? *Revue internationale de droit économique*, 3, 351–369.

Rose, C. (1986). The comedy of the commons: Custom, commerce, and inherently public property. *University of Chicago Law Review*, 53, 711–713.

Sagot Duvauroux, D. (2002). La propriété intellectuelle c'est le vol. In *Les majorats littéraires*, Proudhon, J. (ed.). Les Presses du réel, Paris.

Salaün, J.M. (2012). *VU, LU, SU, les architectes de l'information face à l'oligopole du web*. La Découverte, Paris.

Samuelson, P. (2003). Mapping the digital public domain: Threats and opportunities. *Berkeley Law Scholarship Repository*, 66(1), 6.

Scholtz, T. and Schneider, N. (2016). *Ours to Hack and to Own: The Rise of Platform Cooperativism: A New Vision for the Future of Work and a Fairer Internet*. OR Books, New York.

Shapiro, C. and Varian, A. (1999). *L'économie de l'information*. De Boeck, Brussels.

Stiglitz, J. (2006). *Un autre monde. Contre le fanatisme du marché*. Fayard, Paris.

Suber, P. (2016). *Qu'est-ce que l'accès ouvert ?* OpenEdition Press, Marseille [Online]. Available at: http://books.openedition.org/oep/1600.

Tirole, J. (2016). *Économie du bien commun*. PUF, Paris.

Vanholsbeeck, M. (2017). La notion de science ouverte dans l'espace européen de la recherche. *Revue française des sciences de l'information et de la communication*, 11(2017) [Online]. Available at: http://journals.openedition.org/rfsic/3241.

Weinstein, O. (2015). Comment se construisent les communs. Questions à partir d'Ostrom. In *Le retour des communs*, Coriat, B. (ed.). Les liens qui libèrent, Paris.

# Index

Other titles from

in

Information Systems, Web and Pervasive Computing

## 2021

EL ASSAD Safwan, BARBA Dominique
*Digital Communications 1: Fundamentals and Techniques*
*Digital Communications 2: Directed and Practical Work*

GAUDIN Thierry, MAUREL Marie-Christine, POMEROL Jean-Charles
*Chance, Calculation and Life*

## 2020

CLIQUET Gérard, with the collaboration of BARAY Jérôme
*Location-Based Marketing: Geomarketing and Geolocation*

DE FRÉMINVILLE Marie
*Cybersecurity and Decision Makers: Data Security and Digital Trust*

GEORGE Éric
*Digitalization of Society and Socio-political Issues 2: Digital, Information and Research*

HELALI Saida
*Systems and Network Infrastructure Integration*

LOISEAU Hugo, VENTRE Daniel, ADEN Hartmut
*Cybersecurity in Humanities and Social Sciences: A Research Methods Approach (Cybersecurity Set – Volume 1)*

GHLALA Riadh
*Analytic SQL in SQL Server 2014/2016*

JANIER Mathilde, SAINT-DIZIER Patrick
*Argument Mining: Linguistic Foundations*

SOURIS Marc
*Epidemiology and Geography: Principles, Methods and Tools of Spatial Analysis*

TOUNSI Wiem
*Cyber-Vigilance and Digital Trust: Cyber Security in the Era of Cloud Computing and IoT*

# 2018

ARDUIN Pierre-Emmanuel
*Insider Threats*
*(Advances in Information Systems Set – Volume 10)*

CARMÈS Maryse
*Digital Organizations Manufacturing: Scripts, Performativity and Semiopolitics*
*(Intellectual Technologies Set – Volume 5)*

CARRÉ Dominique, VIDAL Geneviève
*Hyperconnectivity: Economical, Social and Environmental Challenges*
*(Computing and Connected Society Set – Volume 3)*

CHAMOUX Jean-Pierre
*The Digital Era 1: Big Data Stakes*

DOUAY Nicolas
*Urban Planning in the Digital Age*
*(Intellectual Technologies Set – Volume 6)*

FABRE Renaud, BENSOUSSAN Alain
*The Digital Factory for Knowledge: Production and Validation of Scientific Results*

GAUDIN Thierry, LACROIX Dominique, MAUREL Marie-Christine, POMEROL Jean-Charles
*Life Sciences, Information Sciences*

GAYARD Laurent
*Darknet: Geopolitics and Uses*
*(Computing and Connected Society Set – Volume 2)*

IAFRATE Fernando
*Artificial Intelligence and Big Data: The Birth of a New Intelligence*
*(Advances in Information Systems Set – Volume 8)*

LE DEUFF Olivier
*Digital Humanities: History and Development*
*(Intellectual Technologies Set – Volume 4)*

MANDRAN Nadine
*Traceable Human Experiment Design Research: Theoretical Model and Practical Guide*
*(Advances in Information Systems Set – Volume 9)*

PIVERT Olivier
*NoSQL Data Models: Trends and Challenges*

ROCHET Claude
*Smart Cities: Reality or Fiction*

SALEH Imad, AMMI, Mehdi, SZONIECKY Samuel
*Challenges of the Internet of Things: Technology, Use, Ethics*
*(Digital Tools and Uses Set – Volume 7)*

SAUVAGNARGUES Sophie
*Decision-making in Crisis Situations: Research and Innovation for Optimal Training*

SEDKAOUI Soraya
*Data Analytics and Big Data*

SZONIECKY Samuel
*Ecosystems Knowledge: Modeling and Analysis Method for Information and Communication*
*(Digital Tools and Uses Set – Volume 6)*

## 2017

BOUHAÏ Nasreddine, SALEH Imad
*Internet of Things: Evolutions and Innovations*
*(Digital Tools and Uses Set – Volume 4)*

DUONG Véronique
*Baidu SEO: Challenges and Intricacies of Marketing in China*

LESAS Anne-Marie, MIRANDA Serge
*The Art and Science of NFC Programming*
*(Intellectual Technologies Set – Volume 3)*

LIEM André
*Prospective Ergonomics*
*(Human-Machine Interaction Set – Volume 4)*

MARSAULT Xavier
*Eco-generative Design for Early Stages of Architecture*
*(Architecture and Computer Science Set – Volume 1)*

REYES-GARCIA Everardo
*The Image-Interface: Graphical Supports for Visual Information*
*(Digital Tools and Uses Set – Volume 3)*

REYES-GARCIA Everardo, BOUHAÏ Nasreddine
*Designing Interactive Hypermedia Systems*
*(Digital Tools and Uses Set – Volume 2)*

SAÏD Karim, BAHRI KORBI Fadia
*Asymmetric Alliances and Information Systems:Issues and Prospects*
*(Advances in Information Systems Set – Volume 7)*

SZONIECKY Samuel, BOUHAÏ Nasreddine
*Collective Intelligence and Digital Archives: Towards Knowledge Ecosystems*
*(Digital Tools and Uses Set – Volume 1)*

## 2016

BEN CHOUIKHA Mona
*Organizational Design for Knowledge Management*

BERTOLO David
*Interactions on Digital Tablets in the Context of 3D Geometry Learning*
*(Human-Machine Interaction Set – Volume 2)*

BOUVARD Patricia, SUZANNE Hervé
*Collective Intelligence Development in Business*

EL FALLAH SEGHROUCHNI Amal, ISHIKAWA Fuyuki, HÉRAULT Laurent, TOKUDA Hideyuki
*Enablers for Smart Cities*

FABRE Renaud, in collaboration with MESSERSCHMIDT-MARIET Quentin, HOLVOET Margot
*New Challenges for Knowledge*

GAUDIELLO Ilaria, ZIBETTI Elisabetta
*Learning Robotics, with Robotics, by Robotics*
*(Human-Machine Interaction Set – Volume 3)*

HENROTIN Joseph
*The Art of War in the Network Age*
*(Intellectual Technologies Set – Volume 1)*

KITAJIMA Munéo
*Memory and Action Selection in Human–Machine Interaction*
*(Human–Machine Interaction Set – Volume 1)*

LAGRAÑA Fernando
*E-mail and Behavioral Changes: Uses and Misuses of Electronic Communications*

LEIGNEL Jean-Louis, UNGARO Thierry, STAAR Adrien
*Digital Transformation*
*(Advances in Information Systems Set – Volume 6)*

NOYER Jean-Max
*Transformation of Collective Intelligences*
*(Intellectual Technologies Set – Volume 2)*

VENTRE Daniel
*Information Warfare – $2^{nd}$ edition*

VITALIS André
*The Uncertain Digital Revolution*
*(Computing and Connected Society Set – Volume 1)*

# 2015

ARDUIN Pierre-Emmanuel, GRUNDSTEIN Michel, ROSENTHAL-SABROUX Camille
*Information and Knowledge System*
*(Advances in Information Systems Set – Volume 2)*

BÉRANGER Jérôme
*Medical Information Systems Ethics*

BRONNER Gérald
*Belief and Misbelief Asymmetry on the Internet*

IAFRATE Fernando
*From Big Data to Smart Data*
*(Advances in Information Systems Set – Volume 1)*

KRICHEN Saoussen, BEN JOUIDA Sihem
*Supply Chain Management and its Applications in Computer Science*

NEGRE Elsa
*Information and Recommender Systems*
*(Advances in Information Systems Set – Volume 4)*

POMEROL Jean-Charles, EPELBOIN Yves, THOURY Claire
*MOOCs*

SALLES Maryse
*Decision-Making and the Information System*
*(Advances in Information Systems Set – Volume 3)*

SAMARA Tarek
*ERP and Information Systems: Integration or Disintegration*
*(Advances in Information Systems Set – Volume 5)*

## 2014

DINET Jérôme
*Information Retrieval in Digital Environments*

HÉNO Raphaële, CHANDELIER Laure
*3D Modeling of Buildings: Outstanding Sites*

KEMBELLEC Gérald, CHARTRON Ghislaine, SALEH Imad
*Recommender Systems*

MATHIAN Hélène, SANDERS Lena
*Spatio-temporal Approaches: Geographic Objects and Change Process*

PLANTIN Jean-Christophe
*Participatory Mapping*

VENTRE Daniel
*Chinese Cybersecurity and Defense*

## 2013

BERNIK Igor
*Cybercrime and Cyberwarfare*

CAPET Philippe, DELAVALLADE Thomas
*Information Evaluation*

LEBRATY Jean-Fabrice, LOBRE-LEBRATY Katia
*Crowdsourcing: One Step Beyond*

SALLABERRY Christian
*Geographical Information Retrieval in Textual Corpora*

## 2012

BUCHER Bénédicte, LE BER Florence
*Innovative Software Development in GIS*

GAUSSIER Eric, YVON François
*Textual Information Access*

STOCKINGER Peter
*Audiovisual Archives: Digital Text and Discourse Analysis*

VENTRE Daniel
*Cyber Conflict*

## 2011

BANOS Arnaud, THÉVENIN Thomas
*Geographical Information and Urban Transport Systems*

DAUPHINÉ André
*Fractal Geography*

LEMBERGER Pirmin, MOREL Mederic
*Managing Complexity of Information Systems*

STOCKINGER Peter
*Introduction to Audiovisual Archives*

STOCKINGER Peter
*Digital Audiovisual Archives*

VENTRE Daniel
*Cyberwar and Information Warfare*

## 2010

BONNET Pierre
*Enterprise Data Governance*

BRUNET Roger
*Sustainable Geography*

Printed and bound by CPI Group (UK) Ltd, Croydon, CR0 4YY